D1193714

F[GHTING
THE GOOD FIGHT
For Municipal Wireless

Craig Settles

ISBN-13: 978-1-58776-836-1
ISBN-10: 1-58776-836-4

Manufactured in the United States of America

1 2 3 4 5 6 7 8 9 10 NetPub 0 9 8 7 6 5

675 Dutchess Turnpike, Poughkeepsie, NY 12603
www.hudsonhousepub.com   (800) 724-1100

Dedicated to the memory of my uncle, Emmett A. Murphy, former Chief of the Bureau of Weights & Measures, Philadelphia Department of License and Inspections. He fought the good fight on behalf of the average citizen for whom he believed the scales of commerce should always be fairly balanced.

Another Good Book For Your Project Team!

Pilots to Profits: Getting in Sync with the Mobile Mandate

Rip Gerber
Craig Settles

ISBN: 1-58776-835-6
Soft Cover
234 Pages

Benefit from the 20/20 hindsight of over 50 executives, business managers and IT professionals whose detailed roundtable discussions present their experiences, insights and useful tips for implementing applications with measurable ROI. Planners and implementers will all benefit from these lessons.

Pilots to Profits helps you define, pilot and effectively deploy mobile and wireless applications. Security, change management, budgeting, pilot project logistics, build-or-buy decisions, and aligning business objectives with IT initiatives are just some of the topics addressed.

Starting with "why is now a good time to seriously consider mobile and wireless," this book moves readers through the key strategic and tactical issues that affect the process. Each chapter is designed to be effective by itself so you can read cover to cover, pick and choose chapters based on your pressing needs.

About the authors

Craig Settles, with 18 years experience helping organizations understand how technology can improve their bottom line, punctuates participants' observations with additional points to consider.

Rip Gerber, Chief Marketing Officer for Intellisync, adds critical viewpoints from the vendor side of the deployment equation. Any major project involves multiple-vendor participation which brings with it multiple opinions, agendas and solutions.

Together, their industry, technology consulting and strategic planning expertise helps you avoid many pitfalls and setbacks. Whether you're starting your first mobile or wireless deployment, or you already have some battle scars, "Pilots to Profits" will be your constant companion.

Acknowledgements

First, I would like to acknowledge and thank the Wireless Internet Institute, through whose help I was able to get the direction and access to people needed to quickly complete this project. Also, many thanks to Dianah Neff whose time, assistance and input was equally valuable. Thanks to the people who contributed valuable time for interviews and each have their 15 minutes of fame within these pages. Finally, heartfelt thanks go to all of those citizens who rallied to the cause, made their voices heard and forced the powers that be to exempt Philadelphia from the onerous effects of House Bill 30. Without you, Philadelphia's wireless would not have come this far.

Craig

Table of Contents

Preface

Congratulations Philadelphia! Through words and actions you have lead a national and international discussion on the value of deploying citywide broadband wireless. You have taken a great leadership role, together with Seattle, San Francisco, Austin, Portland and Atlanta, to become one of America's most unwired cities.

There is no right way or wrong way for this journey but you are on your way. The process and decisions you made will potentially help the other 14,428 cities in America, and even more importantly, help urbanized communities find their path to connectivity. This book is an in-depth "How to."

We are evolving into a networked society where connectivity is the key to increasing the quality of how we live, work, play and learn. The economics of providing connectivity are changing with a communications variant of Moore's Law—that bandwidth is becoming more robust and more economical every 18 months. Not so long ago the mantra was that FTTH (Fiber-to-the-Home) made business sense even though it was an expensive undertaking. While wireless can't deliver the bandwidth of fiber and probably never will, it offers ample if not robust bandwidth at costs one-tenth of FTTH and prices are continuing to drop. At these cost benefits, forward-thinking local governments can get truly serious about connecting their constituents more economically than thought possible.

Those of us in the public sector participating in the telecommunications policy reform debate are keenly aware of the 'build-out' issue wherein the communications giants seek only to serve those segments of a community that meet well defined, bottom-line business objectives. For those reading this book from a private sector and not a public policy view let me say that the economics of wireless connectivity can be used to balance these digital divide issues.

Wireless services are especially important in addressing rural and low density connectivity which, in the political scheme of things, is critically important to the majority of the Senate membership representing constituencies with keen rural interests. Wireless technologies offer an economical solution to those areas whose elected officials would like their constituencies offered any type of broadband connectivity. Simply, wireless connectivity has the potential to clear the logjam for national telecommunications policy reform and reduce the digital divide.

The clear winners in municipal provisioning, other than the residents, are the local business owners and *potential* business owners. Wireless connectivity is an attractive tool in municipal economic development and luring higher paying, white collar jobs to any community. New businesses are enticed by cheaper connectivity. Local and regional businesses, especially those with mobile workers, benefit from less expensive and faster mobile connections.Businesses in more rural communities have an option of less expensive broadband in their offices. At the end of the day, workers will be able to do more, do it faster, and do it cheaper.

As Craig points out, there are two choices for local government's deployment of wireless, partnering with providers or doing it yourself. His book documents and describes how Philadelphia made these and other choices. The message for local governments is not necessarily to copy Philadelphia's decisions but to emulate their process. Cities and counties should engage the discussion and decide if connectivity warrants the same level of attention and planning as other critical infrastructure elements such as water, power, sewers and storm drains. If it does, then begin the process of deciding how to get there. This book can help you think through that process.

Good luck Philadelphia and good luck America.

Lori D. Panzino-Tillery

President, National Association of Telecommunications Officers and Advisors (NATOA)

Introduction

Let's get right to the heart of the matter. The decision in Philadelphia to deploy broadband wireless citywide was not a political one, though it obviously impacted many political decisions. It was a sound business decision made by an organization, a major customer of technology, to improve its ability to deliver a better product - government services - to its customers who are the citizens of Philadelphia.

Philadelphia's executive steering committee that was chartered to create a plan to make municipal WiFi work evaluated their technology options as does any prudent business. The committee then selected options that will give them the best performance while delivering the most effective results at the best price point. In fact, they did a more thorough job of analyzing end user technology needs than major corporations spending much more for their deployments.

The resulting frenzy of counterproductive legislation and legislative saber rattling really isn't about political decisions either. It's the disappointing consequence of business decisions made by a tiny group of companies attacking perceived threats to its market position that, in reality, is threatened by its business models which are at odds with the advance of technology.

Fighting the Good Fight is about smart business decision making municipal wireless broadband and mobile government workforce applications. It gives elected officials and managers in local governments, as well as their stakeholders, guidelines for making appropriate wireless plans to meet their specific operational and constituent needs. It gives government CIOs and IT staffs a good picture of vital business issues to address when selecting, then managing, vendors and technology products.

Follow how Philadelphia's project team approached its municipal WiFi decision, did extensive needs analysis, wrote a comprehensive business plan in just three months, launched pilot projects and selected the vendor team to do full deployment. The team withstood the slings and arrows of outrageous controversy that comes from being on the leading edge of muni WiFi deployments, and subsequently they have many lessons for those pursuing this technology.

When you look at the public face of municipal broadband wireless as presented in much of the media, you see political intervention and philosophical discord. You see the needs of the underprivileged pitted against the interests of giant telcos and cable companies. Often you see a race to be "first" in muni Wifi, a fleeting concept at best. What's missing is an in-depth discussion about the business issues.

Fighting the Good Fight starts with an overview of why now is the time for cities and counties to evaluate municipal WiFi. Apart from the representative side, government is a business that takes in revenue to pay employees to deliver services citizens expect. Local governments face increasing challenges that put a strain on financial and human resources as they attempt to achieve their mission. WiFi, when effectively deployed, radically improves how a government does business to the benefit of all of its constituents.

Philadelphia faces intense social and economic challenges that cities of all size face, and the city is meeting these in the conventional ways of many governments. But Mayor John Street felt they needed out-of-the-box thinking to do better. Though the project team had to adjust their thinking several times in the early stages, they never wavered in their belief that new technology can solve old problems.

The next two chapters address the business case for using WiFi and wireless applications for both government and constituent groups that augment the delivery of government services. Local governments need to deploy the WiFi infrastructure to wireless-enable its workforce, then use the resulting return on investment from increased productivity and reduced operating cost as well as creative business partnerships to achieve social objectives.

The $2 million annually that WiFi is expected to save Philadelphia in telecommunication costs for its 2000 mobile workers and 300 remote offices makes for a solid business case. But then there are wireless applications within departments such as the Licenses & Inspections Department that will each generate their own ROI that should be greater as a result of the citywide WiFi network.

Chapters 4 – 6 address developing a winning project team, building consensus among a diverse assortment of stakeholders representing business, community, education, and other interests, and assessing constituent needs. These three tasks are the foundation which any successful deployment must have. Executing either task poorly leaves you vulnerable to failure, or at least failure to achieve the full benefits that the technology offers.

Pay careful attention here because, as focus groups facilitator Robert McNeil of the Ronin Group says, "You can really go by what Philadelphia did because Philly did a lot of things right." Their approach to picking a project team with the most inclusive representation of constituent groups, building consensus and soliciting feedback was downright obsessive. But it produced good – and unexpected - insights that cities ignore at their peril.

Chapter 7 addresses key issues that define your eventual wireless implementations. Many are technology related and affect workforce applications. However, a major defining issue for the broadband wireless infrastructure is what business model do you employ? Who owns the network? Is access free or paid? Philadelphia chose a "wholesale" model and created a non-profit entity, Wireless Philadelphia, to implement the business plan.

In Chapter 8, you will learn some valuable lessons on acquiring the right technology products or services based on the needs of government workers and constituent groups. There is an incredible array of options, particularly when it comes to wireless applications for mobile workers. You have to do some serious leg work before, during and after deployment to make sure what your users and constituents get is what they need.

Matching technology with mobile workers' needs is relatively easy since they work within a centralized business structure. But Wireless Philadelphia's project team was a little awed when it became apparent how much work must go into community relations to align technology with a crazy quilt of needs, skills, perceptions and expectations among constituent groups. Technology is actually just a part of the solution.

The logistics of deployment are covered in Chapters 9 and 10. I often wonder if the people announcing plans to make their cities wireless to improve operations and attack social issues truly understand the amount of work that lies before them. At one level, IT staffs understand basic technology deployment. But wireless technology fundamentally changes how businesses operate and people conduct business. Wireless deployment isn't just building infrastructure or handing out products. It's herding cats in the most extreme sense of the phrase.

The phase of implementation moving from technology selection to pilot projects to actual deployment requires a blend of business acumen, technology expertise and community relations that you can't address haphazardly. The ability of CIO Dianah Neff and Project Manager Varinia Robinson to talk the talk and walk the walk equally well with businesspeople, techies and average citizens in the neighborhoods is masterful. It's not building the networks that assures their success, it's facilitating what constituents do with the networks.

Two aspects of deployment that governments may not have faced to the extent they will with wireless technology is total cost of ownership (TCO) and return on investment (ROI). There are myriad costs of a wireless deployment beyond the price of hardware, software and services. If you don't prepare for these, down the road a lot of people could be unhappy.

Governments smart enough to act like a business must be good at predicting ROI before deployment and assessing it after deploying. The tricky part is that government differs greatly from many commercial entities in that so much of its ROI are intangible benefits. Saving and transforming lives are not balance sheet line items. Nevertheless, your wireless investment still needs to be balanced against what you achieve with it.

Since there are few citywide muni WiFi deployments completed, TCO and ROI are mainly accurate estimates, best guesses and logical conclusions. Luckily, there are enough benchmarks for mobile workforce applications, radio network deployments by emergency first responders and partial city WiFi deployments to guide the way and fortify confidence levels. Tie your TCO and ROI calculations to department and constituent-level deployments, while heeding advice offered by vendors involved with the Philly project to recommend ways to contain infrasructure costs.

Fighting the Good Fight for Municipal Wireless concludes with an epilogue of forward-looking viewpoints from many of the people involved with the project. Here you get a glimpse of what lies ahead, both for Philadelphia and for you as your municipality begins or continues its muni WiFi efforts.

This book helps you understand what Philadelphia did, why they did it and how similar wireless initiatives can improve local government and the lives of its citizens. Imitation may be the best form of flattery, but you don't want to imitate the project team's decisions, you want to imitate the process by which they came to those decisions.

Municipal broadband wireless has an unprecedented opportunity to change the way in which local governments manage the people's business, and in an incredibly short period of time. It may not be the ideal technology for every government or constituent group. But everyone owes it to their community to evaluate the potential, to explore the art of the possible.

Chapter 1

Why now is the time for municipal broadband wireless

"As mayors, we're on the ground level of dealing with the problems and challenges of local communities. Everybody knows where you live. People will come up to your door and knock on it and expect to talk to you!"

In three short sentences, Mayor John F. Street of Philadelphia crystallizes why it is that mayors everywhere need to seriously evaluate bringing municipal broadband wireless (muni WiFi) to their constituents. There are serious challenges that face our cities of all sizes, and the responsibility for addressing these fall squarely on the shoulders of local elected officials. Many have looked at the problems of today and the technology of tomorrow, and decided that muni WiFi holds the key to overcoming many of these challenges.

Fighting the Good Fight tells Philadelphia's story of how they moved from idea to the actual launch of a major technology initiative to reshape how the city delivers services and additional benefits to its many constituents. This book does it in a way that provides business guidelines to help municipal and county elected officials, government managers, community leaders and businesses develop a plan for using broadband wireless to improve their cities.

While Philadelphia's decision to blanket the city with broadband wireless had political impact, as proven by the resulting whirlwind of media coverage and legislative action, this was a business operations decision. The problem is that the technology's business role was quickly superseded by politics, due in part to Philadelphia being the largest city in 2004 to announce their intent for a full city deployment. Several small cities had decided to deploy muni WiFi, but here was a major metropolis on the verge of dramatically altering the communications world as we know it.

Not only that, the primary objective of the initiative was to attack two substantial ills – the lack of technology access by underserved communities, and economic revitalization of blighted urban areas. Even if there had not been the intense lobbying backlash by incumbent telecom and cable companies, the enormity of the problems being addressed guaranteed widespread attention. A leading city was committing to aggressively wield technology similar to the sword of St. George to slay two huge dragons besieging their realm. Every politician understands the value to their careers and to their legacy of slaying dragons, so many added their voices to the fray.

By viewing Philly's municipal broadband initiative through the lens of responsible process management, as this book does, you learn valuable lessons that lead to better decisions. Local governments are in the business of using their human, financial and physical assets to deliver services that their customers (citizens) pay for with their taxes. Many municipalities are multi-million, if not multi-billion dollar business operations. The more efficiently these operations are run, the greater social and economic good a municipality does. As such, consider muni WiFi a business technology tool that improves use of your major assets while producing profound, positive economic and social benefits. That is, if you manage deployment correctly.

Fighting the Good Fight is not a how-to book that gives you all the answers involving the deployment of muni WiFi. Rather, it presents key questions you need to ask in order to determine how or if you should use municipal broadband to meet community needs. Some cities believe municipal broadband can revolutionize how they do business and lower their cost of doing business. Others are leery of the claims. Most governments are just not sure, but they are curious. The experiences, insights and tips in these chapters bring clarity to a process for which there are few ground rules as of yet, but the promised rewards are great.

This chapter starts moving past some of the political rhetoric to put you on the path to making choices that are right for your specific community. It tackles the question "why is now the time to seriously evaluate making broadband wireless a reality in your city or county?" For those who have started muni WiFi projects, this chapter should reinforce your case for moving forward.

To keep your technology evaluation focused, understand the terms

The discussion about Philadelphia's drive to deploy broadband wireless was sensationalized, distorted and burdened with needless political drama in its early days that still affects discussions in other cities and in the media over a year later. A factor contributing to the drama, and to the question of whether cities should even consider having an active role in bringing the technology to citizens, was that some people didn't understanding key technology and business terms important to the discussion.

The upside of all the attention is the huge national and international spotlight it put on wireless broadband technology and the value that it offers individuals, businesses and governments of all types. More than the combined advertising force of all the vendors involved in providing broadband products and services, this continuous stream of media coverage put muni WiFi on the radar of a broad range of politicians, business people and average citizens.

The challenge to these city leaders pushing muni WiFi plans forward is how to focus various constituencies and stakeholders on a common vision when advocating technological, political and business issues that are quite complex. Four bedrock technology and business terms around which there should have been greater clarity from day one in Philly are "wireless", "high speed", "economics" and "business model". Understanding these in the context of the needs of both governments and citizens illuminates the urgency which drives cities and counties to act sooner rather than later, plus helps you keep the focus on deciding what's best for your communities.

Wireless

First, wireless means more than cellular phones and second, the technology underlying wireless hotspots (WiFi) offers far greater value than enabling pit stops for quick Internet hookups at cafes and libraries. Wireless in the municipal broadband discussion is the transfer without wires of all types of digital data (text, forms, photos and diagrams, video, voice, etc.) between the Internet, computer networks, people and even inanimate objects.

In the wired world, people get connected to information by plugging a telephone cord or cable from their desktop and laptop computers into a wall jack or centralized server. These connections lead to e-mail, databases, software applications, the Internet, services and other people in different locations using computers similarly connected.

With WiFi – wireless fidelity - you're enabling the two-way transfer of e-mail and data without cords or cables between radio frequency (RF) transmitters called access points and computers, similar to how a TV tower transmits images to the TV in your home. Some transmitters are powerful enough to beam data from blocks away to your computer, and others need to be within a few yards. Once you "cut" the wire connection, the impact is huge. And herein lies one stumbling block. The concept of being wireless is so simple, some people can't envision the powerful potential impact.

Once you're wireless, it doesn't matter where you work. Indoors, outdoors, underground or many stories above ground. As long as your computing device can find a transmitter, the world of information is literally at your finger tips. What tech savvy government officials such as Mayor Street have known for several years is that once you enable mobile workers to access and input data from anywhere, their productivity increases can be quite awe-inspiring. The next chapter describes in detail how this translates into huge cost savings at a time when city governments are being forced to shoulder greater responsibilities for its citizens but with stagnate or shrinking budgets.

With the maturity of WiFi in 2004, these tech savvy officials saw two additional benefits, also of great significance. WiFi is inexpensive and it's fast. When you use it in office buildings, it's a better alternative than landline connections. It's less expensive to connect employees' desktops with wireless connections rather than installing and yanking out wiring every time employees come and go, whenever you expand or move offices, etc. More germane to Philly's social and economic development needs, it's cheaper to connect any individual from home to the Net at faster speeds using WiFi than it is for any of the landline options.

When city government folks in Philadelphia assessed where wireless had evolved to by 2004 in terms of mobile devices, infrastructure and so forth, as well as where the technology was going, they had a collective "eureka" moment. WiFi is in practically every computer and laptop produced, it will be in new models of PDAs, smartphones and cell phones, and you can get adapter cards for under $100 to WiFi-enable many older computers. New and planned advances in WiFi technology make it possible to turn large geographical areas into massive hotspots. It became easy for these forward thinkers to see that wireless was the answer to some fairly pressing challenges facing the city.

High speed (synonymous with broadband)

What seems to get lost in political debates is the relationship between the speed of Internet access and its impact on individuals' and businesses' ability to be productive and financially prosperous. High speed access is viewed as a luxury for the home and for well-to-do executives or students with laptops. "Why spend tax dollars to build broadband networks when poor people already get dial up, plus they don't have jobs to be able to afford laptops? Why not spend the money for something better like improving education?"

The reality of today's world is that almost everything you need to do, from improve your job or business condition to improve your health and living conditions is facilitated by the Internet. Dial up connection to most Internet resources is slow because many Web pages and applications are designed in a way that eats up bandwidth. For a business, the difference between dial up and high speed connections for each information search or page access may be two or three minutes rather than seconds to load a Web page, or 30 minutes for a file-download versus a minute. In the big picture view of the 50-hour work week, these wasted minutes translate into hours. In terms of productive use of technology, businesses without high speed are at a competitive disadvantage against companies that get more work done faster with high speed.

Putting the speed issue in human terms, Intel's Paul Butcher, Marketing Manager of the State and Local Governments, America's Marketing Group, described how his son prepared for a debate class, and was able to search the Web using broadband access and find 40 arguments for and 40 against a particular issue in a matter of minutes. "Another child with just dial up or no connection at all would spend hours researching and may not find half as much information. Who is going to do better in class?"

For Philadelphia, with its high unemployment, great educational disparity between rich and low income students, large numbers of disadvantaged businesses and a mandate from the Mayor to turn this around, the issue of speed is critical. If Internet access and the speed of that access is crucial to personal and business success, which it surely is, then WiFi's ability to address these issues makes the technology a key part of the public policy decision.

Opponents tried to block Philadelphia's muni WiFi efforts by saying the high speed provided by the private sector is sufficient. But if you look at the rest of the world, the U.S. adoption rate of broadband has slipped from third place in the world to 16th. The Community Broadband Coalition, a diverse group of businesses, special interest groups and local government organizations, recently sent a letter to Congress urging the passage of a community Internet bill to prevent incumbent telcos from blocking cities' broadband initiatives. The coalition's letter said, "Only 30 percent of U.S. households subscribe to broadband services, a reflection of high prices, too few choices, and unavailability of attractive services." Given that stark reality, Philly continued to push forward.

Speed (or the lack thereof) also affects government business operations, which isn't addressed enough in media coverage. This too was factored into the city's decision about whether 2004 was the time to push for WiFi. For mobile workers, the typical wireless data access happens over the same network that carries cell phone voice calls. Not designed for data traffic, the basic cellular networks are maybe only twice as fast as dial up connections. Faster networks being deployed by cell phone carriers do speed up the connection, and are advertised to work at average broadband speed.

But here's the problem. Research by Philadelphia quickly revealed that their high speed options other than WiFi were not uniformly fast. High speed carrier services are advertised as providing a range of between 400 bps and 700 bps. However, that's only the download speed. Upload speed is maybe 50% faster than dial up - 60-80 bps. That's a pretty dramatic range, plus you're never sure that the speed will be consistent in the same location even at different times on the same day.

What's more, what the U.S. passes off as broadband speed is woefully low when compared to other countries. Rather than motivate carriers in the U.S. to catch up, the FCC defines broadband as starting at 200 kbps of data speed in one direction. In parts of Europe and Asia broadband is as fast as 20 megabits per second and steadily improving.

Philadelphia's WiFi infrastructure is being designed to deliver an average of 1 megabit per second in both directions. Other cities can match this speed depending on how they build out the infrastructure, and as WiFi evolves it will offer the capacity to move closer to international broadband speeds. Because WiFi is based on standards and is backward compatible, cities will be able to migrate from current to faster versions of WiFi without having to replace the existing infrastructure.

The business impact that speed has is in the type of mobile workforce applications that you deploy. So many articles about municipal broadband focus on providing low cost high speed access for general citizens and miss the point that providing uniformly high rates of data transfer speeds gives your mobile employees much greater computing power. They're able to process forms, access schematics, track hundreds of physical assets, search massive databases, video conference and perform other bandwidth-intensive activities while still in the field. This capability only gets better as WiFi capabilities increase.

As you look at the technical aspects of speed, you will find that speed is directly tied to costs, which then brings economics into the decision-making process. Since economic issues are complex and politically sensitive, it can be difficult to advance the case for municipal broadband when public discussion and political debates become clouded by opponents' rhetoric, such as market forces will lead to lower prices.

Economics

Let's see what market forces have done for broadband prices. High speed data packages from wireless carriers for mobile users can range from $60 - $80. Broadband such as DSL or cable in the home can cost $30 - $50 per month for those who can get it, plus residents get locked into lengthy contracts by some providers. Broadband for business or government offices costs an additional $25 or more per month. Even when you see promotions for low-cost broadband from the incumbents, you have to look at the real price.

One company is offering broadband for $15 a month (an offer conspicuous by its arrival after the push for muni WiFi picked up speed). However, the offer requires that customers take the carrier's telephone and long distance services to qualify. The rate goes up to $37.95 after the trial period and there is a fee if customers cancel within a 12-month period. This offer is only for fixed (indoor) wireless. If you want high speed wireless data services for outdoor use, that will cost another $60 per month. As carriers roll out faster networks, these likely will cost more in air time and require expensive add-ons to mobile devices or expensive new devices capable of working with the networks.

To understand the economics driving Philadelphia's decision to pursue WiFi and the incumbents' counter attack, consider these three points. First, the economic reality of $60-to-$80/month per worker to make mobile city workers more efficient is a fairly large hit to take when you multiply this times the two thousand workers that Philadelphia has. On the flip side, the economic reality for companies profiting from these prices is that any threat to these fees is anti-business. It's anti-their-business, though probably pro-business for every other company in town.

Second, many small businesses (1- 10 employees), economically disadvantaged people and elderly people can't easily afford DSL or cable broadband, plus people in many rural or sparsely populated areas don't have the option to get this service even if they can afford it. Contrary to incumbents' claims, how "affordable" is $50/month or more for individuals and businesses struggling to make ends meet and buy the basics of life? It seems a specious industry argument at best and belies the economic reality of the underserved in our communities. Incumbents have no economic incentive to deliver these services to these groups at a price they can afford.

Third, many businesses are no longer competing just locally or even nationally. They're competing globally and the Internet is their key to success in this new world order. If the leading businesses in your community are looking to re-locate to other areas that provide more affordable - or any - sophisticated broadband access, the economics of your city losing those jobs and tax revenues call for dramatic counter-measures. If enticing new businesses to come to your city and getting skilled graduating students to stay are vital to economic growth, as is the case in Philly, then making your communication infrastructure competitive with other high-growth areas is crucial.

Against this economic backdrop was WiFi. For a few million dollars, the city could own the network, equip its mobile workforce and remote offices with less expensive, faster high speed access, attack digital divide issues and encourage economic development. For them, the economics made sense, as it did for anyone who appreciates a more fiscally responsible government.

Conversely, telephone companies, wireless carriers and cable companies suddenly saw the specter of lost market share because once people realize what real affordability is, they will want to tap into it. These business interests resorted to fighting this initiative and preemptively attacking similar initiatives by other cities at the state and federal legislative levels.

Business model

The other root cause of the on-going dustup around Philadelphia is the question of who owns this business of municipal broadband, typically phrased as "cities are getting into a business where they don't belong, at taxpayer expense, no less." Philly's original intent in the summer of 2004 was to own and manage the network infrastructure and service. As the project team gathered feedback from citizens in the fall and also got swept up in a major state legislative battle, they re-thought the business model. Yet to this day, much of the opposition within the city starts with "they shouldn't be in the broadband business."

In early 2005, Philadelphia considered a second, "wholesale" business model in which they facilitated the set up of a 501(c)3 nonprofit, independently operating corporation named Wireless Philadelphia. The goal was to raise money as any other business does, though not from taxpayer funds, have a company come in to build the network and then select a service provider to sell access to citizens and even other providers. The lead wholesale provider would be contractually obligated to re-sell the service to low income residents for about $10 and in the $20 price range for others. Wireless Philadelphia would coordinate projects aimed at using the network for economic development activities. This is a standard government/private partnership that has been the staple of major government initiatives for decades.

In the fall of 2005, a vendor team headed by Earthlink won the RFP bid by counter proposing that Earthlink will fund the design, build out and management of the physical infrastructure. They also will sell and service customer access accounts, and lease access to other service providers as well as to the city government and Wireless Philadelphia.

Understand this about business models. There are various options that each city has and some definitely do not put the city in the broadband business nor dip into tax dollars. As more governments get involved with muni WiFi, and as deployments finally go live, variations of these models are likely to emerge and some models will lose favor. Your city should select whichever one is best after weighing each option against the needs of your government, citizens and the political circumstances you face, then clearly articulate the particulars of that model so everyone understands what you're doing.

Philadelphia CIO Dianah Neff defines some of your options. "There are public/business models. There are private consortiums. We had a couple that bid on our project. Government moves out of the way; they come in and put up the network. There is the cooperative-wholesale model which is what Philadelphia selected, and that you're hearing about in a number of other communities. Actually, we're a hybrid of this, a nonprofit corporation created by a municipal government that is entering into a cooperative wholesale model through a public-private partnership. Then you have the public utility and authority option. Chaska, MN, and Scottsburg, IN are classic examples of that, as well as Adel, GA. These communities own and operate their own utility and they implement WiFi as part of these utility operations."

Work with vendors and providers to think outside the box to create options. After a year of intense debate about the viability of Philly's original model, Earthlink's offer took people by surprise, and quickly changed how other cities think about implementing their plans. This is a good choice for Philadelphia, but if everyone had put blinders on and looked at only the option in their RFP, they might have let their best option slip by. One development to note is that a number of cities are following Philly's lead and setting up a not-for-profit entity to give them greater flexibility in dealing with the ever-changing political, business and technology landscapes that affect the business model.

Three objectives factored into Philadelphia's decision to move forward

Now that you have a basic understanding of some of the key terms that are critical to keeping the discussions you have about municipal broadband properly focused, let's address setting the business and political objectives for municipal broadband. To answer the question "is now the time to pursue muni WiFi?" determine if there are immediate needs that the technology is sufficiently capable of meeting at a price (financial, political and otherwise) that is affordable to everyone concerned. Meeting these needs becomes the main objectives driving your decisions throughout the evaluation and implementation process.

Greg Richardson is Managing Partner of Civitium, the management and technology consulting firm that has helped Philadelphia since its start down the path to municipal broadband. Richardson has been involved with the planning efforts of cities of various sizes throughout the country. He finds that the interest in, and need for, wireless falls into three categories.

"Most cities are doing this for 1) social reasons, which is common in cities wanting to address issues such as the digital divide; 2) economic development, which is more intense in areas that can't foster economic development and bring new companies to the city or region unless they have a better communication infrastructure than what the private sector is motivated to provide; or 3) to increased efficiency in government while reducing high telecom expenses that they might be paying to private sector companies."

Philadelphia focused on the first two factors when setting their main objectives for muni WiFi, which Wireless Philadelphia addresses, and the third will be addressed by the city's department commissioners and may have a slightly lower profile. This decision made sense given the city's pressing priorities. I believe that many other cities should tackle the third category first because this holds the key for financial success that can fuel you efforts to meet the other two needs. Which approach is most appropriate for your city or county? Well, your city has to be the judge. I address business operations here first and then the remaining categories.

Municipal broadband improves business operations

Depending on your politics, the thought of governments running a business operation may or may not raise your blood pressure. However, the reality is that even if municipalities shouldn't run businesses, they should be run more like a business.

As I mentioned earlier, city governments (except the legislative side) operate similarly to commercial entities. They collect taxes and fees to hire and manage employees who produce and deliver a product (services) to customers who are the residents, business owners, tourists and other people traveling into town for various reasons. The "profit" of a city is its ability to deliver the most effective services to the greatest numbers of customers as quickly and efficiently as possible.

Many cities' workers are mobile a great deal of the time, both within and between offices and on the street. Over 2000 employees work the streets of Philadelphia – parking meter readers, building inspectors, public safety personnel, social service workers, etc. Beyond City Hall there are several hundred offices and facilities, including fire and police stations, scattered around the city with office and mobile workers who need to stay connected to the city's operations.

Forget about selling wireless access services for a minute. If you look at what the 650 square-mile city of Houston, TX is planning, you realize that many cities can justify the investment in deploying broadband wireless just with the increase in profitability from improving the efficiency and productivity of its mobile workers. Houston justifies its investment on the basis of revamping its parking meter reading and collection operation.

Houston has 2,000 parking meters in downtown and the surrounding areas, so parking already generates huge revenues. The city is doubling the number of meters, but all 4,000 will be new "smart" meters. Drivers will be able to walk up to an electronic station on the block, push the number for their respective parking spaces and make a payment in any form – coins, dollar bills, credit/debit cards. All of the meter stations wirelessly communicate payment and time data via the Internet back to the main office.

The city will streamline meter management and increase revenue. The number of tickets issued will increase because the stations tells meter readers on their PDAs where the spaces are that people haven't paid for, significantly speeding up workers' efforts so the city earns more from the same size workforce. In addition, every meter doesn't have to be checked as often because there's fewer coins to begin with given the multi-payment option, and the electronic system tells workers exactly which ones have coins. The people who typically walk around doing this can be dispatched in a vehicle where and when needed.

What's truly staggering, though, is the monthly cost savings WiFi produces for operating the meter system. Wireless access to make smart meters work would cost $20 per meter per month for CDPD cellular data service. That's $80,000 each month! With WiFi, there are no recurring costs. The city estimates that they will make their money back for the entire citywide build out in eight months. If you look at the money that is saved beyond the 8th month, the economic and social services benefits for any other uses for wireless that the city thinks of can be justified by the improved parking meter operation.

Dianah estimates that "we can save as much as $2 million annually starting in the third year with the lower cost broadband alternative. Right now we pay for cell wireless for a number of field workers to access data, such as those in public safety who need this access in the vehicles and building inspectors, health inspectors and social workers. This is $70 per employee per month. Because of the costs we have limited the number of people who have access."

When her team evaluated the city's workforce, they realized that the city either has to bring in a whole lot more people to handle mobile workers' increased workload, or use WiFi technology. Instead of having to pay $70 per worker each month for access, the city will pay $20 per month by leasing the service from Wireless Philadelphia. They also can drop 50 – 60% of the cost that they pay today for T1 lines in the city's remote facilities, such as firestations or social service offices. "Even in worse case scenario we will save $1 million by providing WiFi access to mobile workers and remote offices," Dianah states.

The benefits of thinking like a business

Bring all of your departments together to ask them what will happen if we start thinking like a business in terms of broadband wireless? Can we shift the discussion from political hot button issues to one about pragmatic business decisions? More importantly, can we take pressure off of the city's operating budget and maybe even increase revenues by taking a long-term view of what our wireless options are?

Commercial entities use WiFi to wireless-enable the people and physical assets that work primarily on their premises because long-term cost savings and improved efficiencies of wireless access to data and the Internet greatly exceed the investment. It doesn't make economic sense to pay recurring wireless data costs for mobile workers when companies own the assets on which they can place access points and backhaul technology to carry data back and forth between mobile devices and the Internet or intranets.

WiFi is a great option because 1) companies own the network infrastructure where their people work, 2) WiFi uses standards so companies can leverage new technology without buying new infrastructure, 3) they get broadband speeds that are faster than the speed of wireless data access on cellular networks, and 4) they don't pay the monthly per-worker fee for the cellular network. Often businesses recoup their WiFi investment within a year or less in this savings alone.

Shift to municipal governments. Instead of a plant floor or business campus, a city employee's workplace is the city's geographic boundaries. Similar to business facilities, the city owns or can negotiate access to the infrastructure within their boundaries on which WiFi can be deployed. Their employees' productivity and efficiency is greater with broadband rather than slower cellular wireless data service. The cities' budget won't be taking hits from a per-worker or per-asset charge.

When it comes to WiFi-enabled applications, a city's mobile workforce is not much different than the hundreds of workers who are mobile on the premises of a company's huge manufacturing plant, warehouse or shipping dock. If a social worker accesses data from the city's computer network or completes paperwork digitally from a client's home, the time and cost savings are similar to the plant manager accessing their server or completing digital forms while on the plant floor rather than waiting to get back to their desks. Wireless-enabling a parking meter to track and respond to payment activity has revenue-increasing potential the same way that wirelessly tracking products on a store shelf increases revenues in that retail operation.

A number of cities are looking to replicate some version of the Philly government/business partnership. However, within other cities, the government and citizens are deciding that the benefits of the technology are so great that they are willing to pay for it with capital funds. They may or may not directly manage the network after building it, though conventional wisdom is that eventually most cities will leave management to service providers that best understand these issues. Many cities will insist that the providers facilitate the cities' efforts to address community needs. And with either business model, the bottom line is the same. Even if you don't sell one subscription for wireless access to citizens and business, the WiFi investment can deliver significant ROI.

Regardless of the choice your city makes, the business needs of typical city governments make it imperative to at least thoroughly analyze the options. The next chapter gives you steps for doing that analysis. Once you have made a strong economic business operations case for the technology and begin to pursue the technology from that perspective, you put government in a stronger position to tackle community needs.

Addressing the digital divide and other social issues

Elected officials and city government managers who are driven to improve the standard of living for the less fortunate citizens of their communities are pushing for broadband wireless. They believe having access to the technology is necessary for these citizens to stay in the game of a digital economy, and using it effectively is necessary if underserved communities are to have a chance of winning any advances.

Many articles about the use of muni WiFi seem to subtly question whether there is a gap between the haves and the have nots with phrases such as "the alleged digital divide." Here's the reality. The Internet and access to it are resources that require money to build and deliver. Entities that build, support and provide access to its infrastructure want and need to make a profit, so they charge for these resources.

The economics of our world dictate that those who have more money can buy more and better resources. Those with little money have little or no access to these resources. The Net is no exception, and therefore we do indeed have a digital divide. Patricia Renzulli, CIO for the School District of Philadelphia cites surveys that show the range of computer access in homes in the city range from 35% to 95% and this access is directly proportional to the income status of the home.

While businesses do make efforts to close the gap between the fortunate and less fortunate, it is government that has taken the lead in addressing major social or economic divide issues. This is a major reason why many city services have started and why we have government mandated access to inexpensive "lifeline" telephone services and other vital utilities. Increasingly more financial responsibility for providing services is falling on the cities, yet they have to do more with the same financial and human resources.

Governments pioneering municipal wireless broadband's use know they can deliver a number of services faster and more cost effectively if people access them over the Net. Also, residents better help themselves when they can access the Net's other resources. Wireless broadband will give the people who need it the Net access that is vital to their advancement at a price they can afford if the cities take action to ensure low-cost delivery.

The digital divide might be shrinking some here and there, but the need to accelerate closing the gap is increasing. Wireless technology is dramatically changing how people work and how business gets done. As more businesses adopt the technology, and more teenagers drive changes in use of the technology on the consumer side, the less fortunate, the elderly and others on the economic fringes will be left behind in the next flood of advancements.

There doesn't seem to be enough attention paid to the fact that the economic wellbeing of individuals directly impacts business interests. Consider that 82,000 families in Philadelphia live below the poverty line. Both as a consumer base and a potential workforce in a business world increasingly driven by technology, think about the impact of bringing a majority of these people across the divide to become better contributors. At some point, if businesses don't take a greater interest in this issue that affects such large numbers of people, the engines that drive commerce are going to start misfiring, much to the detriment of all. Doesn't that seem to be anti-business?

It is important to note that, if city governments are really serious about making elimination of the digital divide one of their main objectives, then they must be ready to do more than just make the technology available to underserved communities. The worst thing you can do is give someone with few or no technology skills a new technology product or service and then walk away to let them try to figure out on their own how to use it.

Without receiving technology training, a context for its use and content or services that are relevant to their needs, these communities will fail to reap the potential benefits. As a result, they will resent you for dropping unfulfilled promises on their doorsteps and likely not participate in future programs regardless of how well they are structured. To provide these needs, cities are going to need partners.

Those planning the Philadelphia project realized early that community partnerships were going to be as critical to the success of their efforts as the business partnerships with companies such as Earthlink. If broadband wireless is going to impact your citizens at a personal level, you have to quickly identify leaders and organizations within the communities who can help you develop the right content, services and online activities for wireless to enable. Throughout the book you will read about how these partnerships are taking shape in Philly.

Municipal broadband boosts economic development

As great as the need is for individual economic improvement, so too is the need for addressing local business economic issues. Several trends are working against small and large communities. Our national economy has already been moving away from manufacturing and traditional "blue collar" industries to "white collar" knowledge and service industries. But even certain white collar jobs such as programming, telemarketing and customer service are being sent offshore. The U.S. economy is increasingly becoming wrapped into the global economy. Communication in general, and communication over the Net in particular, are key factors in businesses' ability to adapt to these changes.

As a result of these changes and communication's role in meeting them, how well small companies grow and where large companies decide to locate is influenced by the quality of a community's communication infrastructure. Small operations are hard pressed to survive if they depend solely on struggling inner city economies, and are challenged to be able to expand into national or international markets. Affordable broadband wireless access to the Internet is a definite lifeline for businesses to pull themselves into the global marketplace.

Mid-size and large businesses are heavily dependant on the Internet for generating sales, delivering service, managing their workers and running their business operations. Their effective use of the technology determines how well they compete. Businesses in small towns and rural areas are isolated from many resources typically found in large cities, so they really need access to a good Internet communication infrastructure or they will be tempted to move to locations where they can get it.

For the town of Scottsburg, IN, (population 6,000) this scenario unfolded almost overnight. In spite of their requests, the incumbent service operator could not economically justify rolling out last-mile high speed Internet connectivity to the community. As a result, businesses in Scottsburg were contemplating leaving or expanding elsewhere, one of which was a local car dealership that represented 72 jobs and a noticeable contribution to the tax base.

Scottsburg made a $385,000 investment in a broadband-wireless solution that is managed by the town's electric utility. They now provides broadband service to local businesses at half the cost of nearby Louisville. It sets its own rate structure, and two employees do the billing in a local office while utility workers maintain the network. With the new infrastructure, employees have held onto their jobs, at least one business is expanding into e-commerce, the town has maintained its tax base, and the infrastructure is now available to support the expansion of eServices to town employees and citizens.

Another factor influencing companies of all sizes is talented workers. Cities and towns with college communities are finding that keeping these graduates around, or attracting graduates from institutions in other cities, is increasingly difficult if the work and living environment isn't what grads have come to expect. And right now, the average college grad has been weaned on cell phones, mobile text messaging and WiFi everywhere. If businesses want to attract this budding workforce, it helps to operate where the communication infrastructure meets these workers' needs.

For many cities, tourism, conventions and general business travel play important roles in their ability to improve the economic health of all types of local businesses. Wireless communication options play a big role in people's decision to patronize a city, their decisions to return to a city and how they rate a city when talking to friends and colleague.

If your city administration is chartered, as Philadelphia's is, to revitalize the economic infrastructure within and around blighted or distressed neighborhoods, you need to find a cost effective tool to add to your list of options. If one of your missions is to draw new businesses into town, or keep existing ones from leaving, your communication infrastructure is an ace card that you need to be ready to play.

As with using wireless to improve social conditions, if a main objective is to revitalize your local economy, just having easy access available isn't going to cut it. You have to form partnerships with people in the business community who can help you provide content, activities, promotions, etc. that will draw visitors who access the wireless network into those businesses. It's true that conference planners likely will choose a city that has downtown or city-wide WiFi over one that doesn't. But if your wireless access is comparable to competing cities, what will differentiate you is whatever unique amenities are available along with your access.

Looking at the big picture of economic impact, your ability to improve the financial wellbeing of your business community has a direct impact on city revenues, the local job market and the personal economics of the general population. When setting your objectives and evaluating the value of municipal broadband wireless to your area, carefully examine how implementation and use of the technology will influence what businesses stay, grow, come or leave.

Counties have wireless technology needs too

Philadelphia is a county as well as a city, but the city's profile in this discussion shadows county-specific needs that also exist. St. John's County Building Department in Florida, for example, used to have 15 building inspectors completing about 200 inspections a day using a paper-based system that wasted at least an hour a day of each inspector's time. After deploying mobile devices with wireless access capability, 25 people now complete as many as 1000 inspections each day. Less than two times as many employees now do over three times as much work.

"For years this department was viewed by constituents as a big abyss where no one cared, no one liked you and nothing got done. People would've rather taken a beating than go to 'the Buildings Department,'" states Howard White, Deputy Building Official. Today everyone in the field and in the office operates more professionally. They take a lot of pride in their job and being part of an organization that constituents appreciate and respect. "Wireless helps me run this department like a private company. Within bureaucratic restraints, I work diligently to increase our levels of service, be proactive and maintain a competitive edge. If people in government don't recognize the need for this, at some point in time they'll be very disappointed."

Pinellas County Sheriff's Office, also in Florida, typifies how public safety departments can use WiFi to improve operations throughout their multi-building campus for deputies and administrative personnel. They needed a way to deploy two applications that would be able to track inmate movement and electronically record data to inmates' medical records. The system needed to be robust so users could obtain this information at any time in any building within the complex.

Morrow County Emergency Management Center (MCEMC) in Oregon has created what is probably the world's largest WiFi hotspot (1000 square miles). MCEMC needed new communication technology to support its full crew of first responders. They have to move massive volumes if data from cameras used to monitor facilities that house warfare materials, energy production and distribution operations, a natural gas supplier and a nuclear power station. MCEMC also runs an advanced evacuation system with WiFi controls, variable message signs, drop-arms barricades and cameras.

Interview with Mayor John Street

Two central characters in the story of Philadelphia's leadership role in the municipal WiFi movement are Mayor John Street, who had a mission and then a vision, and CIO Dianah Neff who had the skill, drive and also the vision to make things happen. Here, Mayor Street discusses that mission and vision, while setting the stage for the chapters of *Fighting the Good Fight* that follows.

How did you know that the time was right to move forward with citywide wireless?

My CIO, Dianah Neff, is responsible for tracking emerging technologies that can be useful to my administration and the City as a whole. I asked her to prepare a briefing paper that showed how wireless technology could help in my administration's Neighborhood Transformation Initiative [NTI] efforts. NTI, which is a cornerstone of this administration, is a program designed to renew and strengthen entire communities by providing quality housing, clean and safe streets, and vibrant cultural and recreational outlets.

As we looked at citywide WiFi's potential, we saw several significant benefits. This will bring affordable Internet access into low income and minority households. It will help the Philadelphia School District in accomplishing their goal of involving parents more in their children's schooling. From a business perspective, we envisioned this access helping small businesses grow and thrive, and having other benefits for larger companies. This will also help retain more of our college students in the Philadelphia region after they graduate, as well as make Philadelphia a more inviting place for visitors and business travelers to come to.

After considering the possibilities and several more discussions, I asked Ms. Neff to develop and implement a pilot program in LOVE Park, a popular area in downtown Philadelphia, to see what the response from the public would be. We received a strong positive response for this, which led us to then pursue the idea more aggressively.

Did your being a user of the technology influence your decision to make this initiative a priority, or at least make it easier to understand the implications of the decision?

My using technology and believing in the power of technology to transform certainly helped me in my decision to make Wireless Philadelphia a priority for this administration. But I have been a big believer in using technology in this City for the past several years. When I first ran for Mayor in 1999, I said we should use Internet technology to allow customers to do business with the City online instead of building multiple mini-city halls throughout the Philadelphia region.

What are some major challenges a mayor of a large city faces with a technology initiative such as this and how should a mayor address these?

Philadelphia is a big city with many challenges. Anytime you attempt to implement a new major initiative that is risky but holds high potential payback for the citizens, businesses and visitors, you have to proceed thoughtfully. Each community has its unique issues and I was elected to address all of them. So I know you need a program that gets buy-in from the public, helps mitigate the millions of dollars required to build it, and has a strategy to deal with legitimate concerns raised by businesses in the same field.

After you get feedback from the different constituencies, you need to weigh all these risks that potentially come with the project and have the political will to move forward if you believe in the potential benefits. We have done all these things to help this initiative to progress.

What steps went into selecting the people to be on the Executive Committee? How do you make a group such as this inclusive enough without becoming unwieldy?

Based on my twenty-five years of public service, I know who the major stakeholder groups are in this City. We collaborated with the right people from citizen groups, nonprofit agencies, schools, businesses, and the tourism industry. I also sought input from Dianah Neff on the number of people needed to allow us to complete a lot of work in a fairly short time period. We decided that 17 members would be a good size to give us a broad representation of the City and still be able to perform in an expedient manner.

I appointed key community stakeholders to those positions on the Wireless Philadelphia Executive Committee we created, and charged them with studying the needs and recommending solutions that would be sustainable and cost-neutral to the City. In five months, they delivered a good business plan to me for review and approval. I was very happy with the business model which the Committee proposed, and we've adopted, because there will be no cost to the city since no city taxpayer dollars will be used to fund this wireless network.

What tips do you have for other mayors and city officials for building consensus among organizations and constituencies with different issues, priorities, etc. of the project?

Transparency of the process is most important. Discuss wireless issues publicly. If you choose the business model that Philadelphia did, appoint a non-profit entity's board publicly. If the need then exists to solicit and find a vendor to contract with to build and operate your city's wireless network, make sure the non profit entity puts out an RFP and advertises for a vendor through an open, public, competitive process. Conducting this initiative transparently and in the public this way will go a long way towards helping to build consensus among different groups and constituencies. Everyone can see what the city is doing and comment on it should they so choose.

Do you see departments within the city such as L & I, the Department of Human Services and public safety organizations taping into Wireless Philadelphia? If so, what kinds of financial benefits do you see coming to the city government?

The City's Managing Director and various departments have given me positive input concerning the Wireless Philadelphia Initiative. This input, combined with the financial estimates from Ms. Neff, leads me to believe that all departments with personnel who need access to city data from the field will benefit.

One pilot program was launched with the Department of Licenses and Inspections. The agency's Dangerous Buildings Inspectors equipped with mobile devices demonstrated that the City could save as much as two hours per day per inspector by having wireless access in the field. This eliminates the need for them to drive back to the office for new assignments and to file reports. We have estimates that by year three the City could save as much as $2 million annually just in reduced telecommunication costs for all of our mobile workers.

For mayors who are just starting to explore the possibility of municipal wireless broadband, how do you describe the pros and cons of using the Philly model of funding deployment versus a city paying for the deployment itself?

The top 45 cities in the country should be able to find business partners that will share in the risk and fund, or help to fund, the project. In mid-size and smaller cities, it may take the city, local businesses or schools to help fund the project. Understand that a wireless solution potentially will be different for each city based on their communities' needs. There are a number of business models that each city will need to evaluate based on their own criteria.

In the final analysis, do your homework and get community involvement. Keep the process open, public, competitive, and fair. Have a careful, well-considered strategy for dealing with the myriad of financial, logistical and political issues that will undoubtedly arise. True change requires political courage, and a willingness to see beyond the daily crises of government to look into the future, envision new realities and then make those realities happen.

Philadelphia has shown that you can be creative in your approach and ensure there is open competition to bring economic stimulus to your community. As a result, the citizens of this great city are poised to be some of the most tech savvy citizens of the United States.

To sum it up

The business case in many cities is clear. There are significant business operation, economic development and social needs that must be addressed by cities and towns of all sizes. As a revenue-generating, service-delivering organization with a demanding customer base, a city or county government is also a customer of technology that will help it address its business needs. WiFi in all of its forms and formats offers a viable tool to meet those needs. As Cole Reinwand, VP, Product Strategy and Marketing, for Earthlink observes, "when new technology like wireless comes along that offers superior economics plus superior performance to enable affordable mobile broadband solutions, cities and private sector companies will consider them."

To give your efforts wings while allaying the criticism that these initiatives attract from certain quarters, do a thorough needs analysis and develop a cost/benefit profile for building out a citywide WiFi infrastructure to improve government operations. If, as in Houston, the business case proves out, the option is there to build the structure with capital funds, city bonds, or whatever makes sense and meets citizen approval. Outsourcing management of the network to competent vendors will be a wise choice for many.

In this scenario, should you decide that the network can benefit social or economic development goals, use it for that in whatever way and through whatever partnership arrangements make financial sense. You own the network. However, it is good politics, sound financial management and great risk reduction if you outsource much of the constituent-facing services and community programs.

Every municipality and county obviously has to do its own needs analysis to determine what operational and/or service-delivery aspects of their organizations and communities can benefit from broadband wireless. The following chapters offer guidelines and recommendations based on the experiences of Philadelphia, other city and county governments and commercial entities that have implemented wireless to help them run a better business operation. The next chapter presents a framework for both assessing the potential financial and operational impact of wireless and laying the foundation for an effective municipal broadband wireless implementation plan.

Chapter 2

Defining the Strategic Goals and Tactics for Broadband Wireless

Chapter 1 presented the three categories of primary objectives that broadband wireless can help you reach. These objectives are what it is you ultimately want the technology to achieve when everything is in place. This chapter lays out a process to help you clarify how to use hardware and software applications, as well as the community and constituent activities facilitated by the WiFi network, to reach your primary objectives.

Articulating a vision is easy. Deploying the network, though challenging in many respects, is relatively easy since vendors do much of the heavy lifting of physical deployment. Putting the right applications, content and activities in place to maximize the capabilities that broadband wireless offers is truly the challenge and one that people take lightly at the city's peril. Everyone involved with the Philadelphia project will tell you that the work involved and the issues you have to address to do this is staggering. But this effort is vital because the network by itself, no matter how fast it runs, is worthless without the right applications and constituent programs that are used appropriately.

Some of the major lessons learned by the Philly team were: assume what people want or need and you'll often be wrong; in diverse cities, many community and constituent groups need different content and programs; finding out what people need requires lots of feet in the street; matching technology with needs is a multi-step process. As for determining what applications are needed by city employees, similar lessons apply, plus you have to deal with departmental politics and understand what technology is already in place that may help, or possibly hinder, your wireless efforts.

To make this an easier and more effective implementation, the process you're about to learn walks you through questions to ask those within your organization, external stakeholders and partners who help you provide services to various constituencies. Some of these questions should be posed to communities directly. As you address these and review what other governments have done with wireless, additional questions will come up.

Look at this as a "task audit." You're going to examine the many tasks your government does that relate to the three primary objectives you want wireless to achieve. The objective is not to give you all of the right answers for your business plan, but help you map out clearer directions for your research, focus groups, pilot projects, business plan and RFP.

This exploration and analysis process helps you build a business case for the technology by establishing strategic and tactical goals for your applications and activities. With these you can more effectively justify the costs for the deployment, and later evaluate the benefits derived from your investment. The process also is a great way to start building a groundswell of support for the initiative from your constituent groups. Those who actively participate in finding their best solutions are the ones most motivated to help you succeed.

The first list of goals you establish will need to be modified as you conduct focus groups (or however you gather feedback from constituents) and implement pilot projects. That's fine. The idea is to get you off on the right foot so you make fewer mistakes as you go along. You will find that change is about the only thing that's guaranteed at this stage in municipal broadband. Adapt and perform.

How the Philadelphia pioneers did it

When Philadelphia started the planning process in mid-2004 to deploy a broadband wireless network, everyone involved knew they were breaking into unchartered territory. It's very difficult to build a business case in the traditional MBA school approach when few cities have done much more than set up a few square miles as a hotspot in downtown business districts. There were a few small cities and towns that had deployed, or were starting to deploy, broadband in large parts of their municipalities. But the needs and challenges to deployment were not on a comparable scale to the nation's fifth largest city.

The core WiFi technology was solid, standardized and broadly deployed. However, a lot of the infrastructure technology that is the layer between end users and the super-high speed connection directly to the Internet was just coming to market in 2004, supplied by a few relatively small vendors. Prior to that, cities were jury-rigging outdoor networks from components built for indoor WiFi deployments. Logistical tasks as divergent as ensuring compatibility and weather proofing were hit or miss. To say that technology roadmaps were untested is truly an understatement.

Nevertheless, with the Mayor's blessing and a clear objective, Dianah Neff and a staff person developed the first proof-of-concept pilot project that went live in June so various constituents could test it and provide feedback. In August, the Mayor appointed the Executive Steering Committee of 17 members to work with Dianah to develop the business plan which they started writing in September and completed in December.

Mayor Street states that "we asked them to tell us how we can create a 135 square mile hotspot, an area where every household can be connected to the Internet at an affordable rate in a way that distinguishes this city. Our children when they come home from public schools can have the Internet right there in their homes. You know, our businesses no longer are just competing with companies in the city and in the region. We have to create an environment where our businesses don't have to worry about whether they can afford to be online because the world is their marketplace."

The committee recruited talent in September from Temple University, LaSalle College and Drexel University to help research what few business models and technology were available. A project manager, Varinia Robinson, was retained to coordinate activities with vendors and other consultants participating in the project, as well as community relations tasks. While research results were coming in, the group launched a series of 20 focus groups from September through October, and then two town meetings, all of which produced a clearer picture of the needs the network would have to address.

In November, the team was sidetracked by a legislative battle at the Pennsylvania capital instigated by the incumbents. The initial result was a bill that prohibited every city in the state from pursuing a broadband service business. In a dramatic 11th hour showdown fueled by 3000 citizens calling, writing and e-mailing the governor, the city forced the legislature to back down and grant Philly an exemption from the bill and a waiver from litigation. Though this allowed them to continue forward, the Committee's planning was influenced by the battle and the terms of the exemption.

The copy of the business plan and the formal RFP that was written in the first quarter of 2005 was announced in early April. As was mandated by the charter to the Executive Committee, the RFP reflected the feedback and research results the Committee gathered. As a result of additional feedback from various groups, changes in the business climate and new developments in technology, the project team added several addendums to the RFP. In addition, the RFP also allowed for alternative technology recommendations and business model recommendations by those responding to the document.

Several pilots went live in parts of the city in the first half of 2005, mainly as additional proof-of-concept deployments from various vendor teams that planned to bid on the RFP. Network testing, including radio frequency (RF) studies were also conducted. The first community pilots were launched in July and August. All of these activities added to the knowledge of Committee members.

The skills of the people on the Committee and those they brought in to assist them were strong in all the right areas, so they were able to pull off a very good strategic and tactical planning effort under adverse circumstances. As an increasing number of cities announce their plans at the beginning of 2006 to pursue their own deployments, there's more experiences, RFPs, talent and technology roadmaps to learn from to help you develop a strategic plan. What is helpful for you at this point is to streamline and focus the initial needs assessment process through what I call the task audit.

The basic foundation of the business case for wireless

The first step of this task audit is to establish the primary objective or objectives for your initiative based on the options presented in Chapter 1. These would be the Mission Statement in the corporate world. Using Philly as an example, their Mission Statement for the initiative (condensed version) is "Wireless Philadelphia aims to strengthen the City's economy and transform Philadelphia neighborhoods by providing wireless Internet access throughout the City." This capsulizes their strong commitment to the primary objectives of closing the digital divide and increasing economic development.

Next, whoever heads up the early-stage drive for muni WiFi needs to identify who within city government and the various constituent groups will participate. They can be department directors, heads of neighborhood associations, presidents of chambers of commerce and others who have a finger on the pulse of their constituents. Preferably these should be people who can maintain an objective viewpoint, have a critical eye for detail, a basic understanding of what the technology is capable of and can get others to easily understand and a commitment to the vision. The first person to champion the project may not be the one who eventually leads it.

Everyone then starts auditing tasks and defining the strategic uses (goals) of broadband that can enable you to achieve your mission. One goal might be reflected in a statement such as "we will achieve our mission by using wireless to improve the flow of economic development information to underserved constituents."

The tactics are the activities carried out to reach the strategic objective, such as creating a city-sponsored job-training portal for residents, or launching a project with corporate partners to provide computers with WiFi cards to low-income individuals. As you define your strategic goals for the project, appropriate tactics – as well as technology – will become apparent as people understand how the various wireless products and services work.

When I speak to commercial organizations, I outline four general strategic operations of a typical company where I believe wireless technology can help them most to save money, make money and run a better business. If they analyze each major task within an operation in depth, asking the right questions, people can put quantifiable values on the benefits of this particular strategic use of the technology. These values become quantifiable goals for what wireless can achieve as well as benchmarks for measuring success after deployment. On the tactical side, I define categories of activities and projects involving wireless and mobile applications that organizations can pursue to reach the strategic objectives you set.

IT staff should read this and participate in as much of the audit as possible to get a better feel for the business and political issues that the network infrastructure and applications must address. This will, or should, help them work more effectively with vendors to guarantee that what the business side of the organization needs, the IT side delivers by way of appropriate products and services.

To help them understand how the technology can impact the various tasks, your project team and others instrumental in the deployment should get an hour briefing on the basics of broadband wireless before they begin the audit. Emphasize technology and mobile devices that are available today. There are vendors and independent consultants that are capable of providing a good overview briefing. Contact some cities that are far along the planning or pilot project stage and they should be able to recommend someone.

Keeping everyone on track

Make sure that everyone clearly understands and always articulates the mission statement. One of the things that the Executive Committee, project team and the Mayor in Philly did very well was consistently hammer on the same message. Regardless to what might have been said about the project in the media or how people felt about the project, there was never any question about what the mission was. The mission statement motivates and unifies everyone who is involved with the project and who benefits from it, plus it helps keep them on track.

As this process of identifying strategic and tactical goals moves beyond the inner circle of the city team, be sure you recruit good people from the associations, non-profits and community groups who will help you carry out its mission. They are referred to in the book as partners or stakeholders, key organizations in the business of providing services and improving life for their respective constituents: businesses, tourists, ethnic groups, students, health professionals, etc.

You may find that the city and your partners may be in agreement with the primary mission, but will develop different strategic goals for the activities they pursue at the community level because each community is different. It's also possible that you and your partners will have different missions. For example, your project team may decide that they want to focus on improving city government operations while community leaders want to focus on economic development. That's ok. Different strategies can be pursued as long as you do effective consensus building with those partners at the outset of the project (see Chapter 5) so you're not working at cross purposes to achieve the mission.

What you're about to read isn't rocket science, and it intentionally lacks the supplemental verbal and numerical details that you find in academic studies, expensive research reports and wonk writings. But, its simplicity makes it easy for non-technical government officials, managers, end users and constituents to define and describe what it is they want wireless to do in terms of meeting government or constituent needs. You can assign or employ resources later to add the detailed analysis reports with extensive number crunching, spreadsheets and pie charts.

Strategic considerations

In terms of assessing the strategic value of the technology, consider these four categories: 1) communicating more effectively and efficiently with existing constituents; 2) enabling constituents to get more effective service and support; 3) improving internal communications and business operations and 4) communicating more effectively with potential constituents.

Read through these next sections once to get a feel for the general strategic issues. Then go back through them a second time to determine which strategic areas are more likely to be a factor in achieving your primary mission. From the beginning, put quantifiable values to your goals. It's not enough to say "we're going to use wireless to help county workers communicate more effectively from the field." Be specific. "We expect broadband wireless to enable each employee to spend two extra hours per day on roadwork through more efficient job dispatch, and by eliminating our paper-based job reporting system."

Communicating less expensively and more effectively with existing constituents

Every city government and organizations partnered with them such as tourism offices, convention centers and so forth incur costs for general communication with constituents, whether they are residents, business people or visitors. There are printing and mailing costs for correspondences, updates, newsletters and the myriad reports and other public documents. You have costs for workers who take calls and handle in-person inquiries. Another expense is the publicity that supports all the various activities and programs hosted by the cities and their partners, plus the resulting phone calls that these activities generate.

The various groups funded or otherwise supported by cities to deliver services to the neighborhoods spend a good portion of their budget on community outreach in the course of alerting residents about programs that are available, where and when. They, as well as state agencies who coordinate services with the cities, all field phone calls and respond to walk-ins seeking information. Don't forget the local business organizations such as the chambers of commerce whose members' interests are often complementary with the cities' economic development efforts. These groups have sizeable communication costs as well.

Along with expenses paid to support all of these communication activities, there are also the "lost opportunity costs," such as not being able to communicate with certain constituents because the organizations don't have enough people, time or money. This is particularly a factor with underserved communities where residents have dropped off the radar screens because they don't have convenient or affordable means to keep the communication lines open. However, the equation holds true for tourists or businesses that you don't have the resources to reach, or reach frequently enough.

The value of wireless as a cost reduction tool can be summed up simply – more often than not it is cheaper to send or access information wirelessly than it is to use print, faxes, telephones, radio, television and other forms of conventional communication. Every hour that can be saved due to municipal wireless is a savings that goes into the business case for the technology. Measurable savings of paper, postage and printing costs attributable to wireless further boost the business case. The ability to show that wireless enables a worker or volunteer to communicate with 15 constituents in an hour instead of 10 drives the business case.

Municipal wireless can create additional quantifiable benefits through your workers' ability to do their jobs more efficiently, or by the technology giving constituents access to information they didn't have before.

Making the business case

To establish goals and create a business case in this strategic area, audit your main communication expenses. How many hours that are being spent visiting neighborhoods can we save if our constituents have access the Internet? What printing or mailing costs can we eliminate if we deploy wireless kiosks or secured wireless-enabled computers to distribute information from the city's Web site? How many more people can we drive into local businesses if we send promotional materials to workers or tourists in our downtown area who are carrying WiFi-enabled mobile devices? Upcoming chapters will point out other examples of how you can save money or do more with your money.

When you do your audit, calculate how much you are currently spending per month or year on conventional communications broken down by city department, community, business organization, other groups or the city or county as a whole. Next, determine approximately how many constituents are being contacted within these groups, or are reaching out to them. Divide the monthly or annual communication costs by the number of constituents to see how much is being spent per constituent. Finally, estimate how much of this cost you can eliminate or significantly reduce with municipal broadband.

In a very simplified calculation of a purely hypothetical situation, a group may find that it's spending an average of $500 per year per constituent to reach 1000 residents without Internet access through direct mail, case workers delivering information to homes and handling in-bound phone calls requesting information. This total amount of $500,000 is determined after factoring in costs for employee time, printing, production, phone lines, etc.

Then they estimate that if these constituents had a basic PC with a WiFi card, the group could pass 60% of all of the information over the Net (not everyone receiving a PC may use it to get your information) so you save $300 in communication costs per citizen. This is a $300,000 savings for reaching 1000 residents. What's more, maybe the employees who don't have to make home visits can now bring several dozen new constituents into the social services program each month, people who would otherwise slip through the cracks.

A similar case can be made for the tourism board that finds they spend $20 per year per tourist to reach 10,000 hotel guests, and maybe this marketing effort drives a collective $50,000 to local retailers (i.e. clothing, souvenirs, books) from the 500 tourists who respond and spend an average of $100 downtown. But a wireless portal that pushes out centralized Web content to tourists from all of the downtown merchants drops the average cost per tourist to $18, while the number of shoppers increases to 1000 when people opt in to get promotional information on their PDAs. You still have to spend the same amount for reaching the 9,000 without mobile devices but only a penny or two to reach mobile devices. The average cost to reach each visitor doesn't drop much, but the revenues generated to retailers doubles.

The numbers you get from your audit are likely to be more detailed than these. How you calculate the expense of doing business or quantify the value of activities you eliminate with wireless will vary from city to city, group to group or even by city department. The what-if scenarios you run can be few or infinite. Often you will find that the financial numbers you get after deploying a pilot are different than what you estimate during an audit. Municipal broadband is new so there's little precedent and no set rules governing the composition of a business case, but there are some procedures that are wise to consider.

1) Isolate the main activities that you and your affiliated organizations are pursuing to communicate to your various constituencies. 2) Assign values of time, money or other resources you are spending in these activities. 3) Get a clear understanding of what wireless can do to improve or eliminate these activities while still enabling

you to achieve your mission. 4) Develop credible estimates of just how much the technology will reduce costs, increase access and streamline efforts. 5) Relate all of these calculations, together with the calculations from the next three strategic areas, to the cost to deploy the technology. Also relate the calculations to the city's primary mission. There may be no great monetary benefit to offset the line-item expense of bringing isolated senior citizens into the mainstream of their communities. But if the city places high intrinsic value on this goal, you may feel the investment is worthwhile.

Let's look at the other strategic options so you can then get a complete picture of how to determine the total potential impact of broadband wireless in your municipality. These five procedures will be applicable to all four of the strategic areas. You may find that you can't pursue all four strategic areas at one time, or your municipality is so small that the lines between the four areas get blurred. No problem. By having these distinct areas, it adds needed structure to the audit process. Adapt it as appropriate to meet your situation.

Enabling constituents to get more effective service and support

The primary product of city government is service. And while giving people easy access to general information such as transit schedules and community news is technically a service, the aspect of service in this second strategic area is helping citizens and visitors get things done, mostly on their own. Filing complaints to get potholes fixed. Processing business tax forms. Getting the living assistance program staff to change services for a parent. Making it possible for clients to submit progress reports to the state welfare-to-work agency without having to spend hours traveling back and forth or sit an hour in a meeting. Creating a wirelessly-accessible system for business travelers to locate and contact an after-hours copy center to place a print job order. Providing self-directed learning and skills development programs for students or adults.

Governments are finding that the more they use the Internet to give control of the service delivery process to its citizens and visitors, the more they reduce costs on the part of service recipients and providers. Also, more people can be served faster and with greater efficiency. As with the previous strategic area, you need to identify those phone, fax, face-to-face and online transactional activities that carry a cost which wireless access can reduce, or whose efficiency the technology can improve. One of the main areas to review in this strategic operation is service delivered to traditionally underserved communities and tourists.

Completing, submitting and tracking government paperwork is the top-of-mind solution that many are familiar with when they think about online "self-help." However, self-help goes many layers deeper when people can do high speed searches of databases or run complex interactive applications from home or wherever they can reach wireless-enabled kiosks or computing devices. Low-income residents need to not only get past hurdles of processing paperwork, they also need to find options and resources that are available to them and be able to use those from the Web. Patricia DeCarlo, Executive Director of Philadelphia's Norris Square Civic Association, observes that "Temple University has all their jobs online, and if you want a job with them you have to go on the Net. But people don't know that."

To do a thorough analysis on the potential impact of broadband wireless, be sure you factor in the numbers of people whose education, economic or living status you can improve through increasing their knowledge and skills. If we link our economic assistance online forms with career development and job search content, can we increase by 10% the number of people who get and keep jobs?

When it comes to tourists and business travelers, the analysis is truly financial and you can take pages from companies' wireless playbook. By how much can we increase foot traffic to our tourist attractions if visitors can make restaurant reservations or book tours from a PDA? What if a convention presenter can send their PowerPoint presentation to a local copy shop while waiting for a shuttle at the airport, and have handouts printed, boxed and ready to go by the time that person arrives downtown? The revenue implication of using WiFi to increase visits to restaurants, retailers, museums and other points of interests, as well as facilitate secure online sales transactions with them, are huge!

To determine the business case in this strategic area, do an audit of all of the service delivery processes which constituents can initiate and even complete on their own if they had wireless access to Web content, online tools and public data. For each process look at the time costs to both constituents and your organization's employees, including travel, meetings and waiting in line or on hold on the phone. Assign dollar values to these where possible, and then determine the financial impact of reducing these costs with wireless.

Giving constituents more control may require moving some city workers' past their comfort zone since access to knowledge, records and internal procedures has long been the purview of government employees. However, so many of them are overworked these days that a good pilot project may motivate them to give you ideas on how you can give even more control to constituents.

One of your audit results may be similar to the following. Each month we have 300 citizens call in requesting applications and help with joining job assistance programs, and another 200 people stop by the office to pick up applications. Each staff person spends an average of 10 minutes per citizen on the phone call while each citizen spends an additional five minutes on hold. Staff spends 20 minutes with each walk-in client. What's more, the office receives 100 calls per month from people seeking the status of their applications or to make updates, which take an average of 10 minutes of staff time per call to resolve.

Collectively, we're spending 125 hours per month handling general requests for information and distributing applications which could be accessed online directly by citizens. What's more, we spend 38 hours per month processing applications received by mail from 150 of the 300 people who call in requesting them. This is a total cost of "x" dollars (163 times the average hourly cost of the employee). If we move 10% - 20% of these job seekers to community centers with wireless-enabled computers, or get the computers into the homes of job seekers, we can trim "y" dollars of costs from the process and re-direct our staff to other important tasks. We also estimate that, by giving job seekers easy access to other Web-based skills development resources, we can possibly increase their odds of retaining employment by 25%.

This audit obviously needs to go beyond addressing the underserved and cost reduction. Similar assessments should be possible for a broad range of city-delivered services to people and businesses in all economic strata. Convention, tourism and business support organizations need to do similar audits to determine all of the various services they provide that drive patrons into business establishments. School administrators should contact the Philadelphia School District to look at the creative learning solutions they implemented that are quite effective, but consume an incredible amount of bandwidth that only high speed technology makes possible.

How much can your visitor's bureau increase traffic into museums and stores if it facilitates the use of mobile devices to map driving directions, or to place orders that are waiting for shoppers when they arrive at a merchant? What is the boost in revenues if chambers of commerce facilitate easy wireless access between their service business members and convention attendees? WiFi-enabled self-service for visitors, particularly after regular business hours, has long-term impact because the positive experiences of these individuals influence many other potential visitors when your guests return to their home towns.

I'm giving you here a simple skeleton of the possible business case you can build. Bring in the appropriate experts and stakeholders to put the meat on the bones.

Improving internal communications and business operations

This third strategic area is where many government organizations will see their greatest return on investment. Using mobile and wireless technology within an organization can lower operating costs, increase employee productivity and improve the effectiveness of working with contractors and other third parties. This audit is also valuable for the non-profit organizations and community groups that align with the city's initiatives since their improved operating efficiency enhances the city's initiatives.

Philadelphia's Department of License and Inspection have already knocked four days off of an emergency violations reporting process because weekend on-call supervisors are completing and submitting the paperwork from their wireless devices as calls come in. Embedding wireless technology into physical assets such as vehicles, smart parking meters and maintenance or public safety equipment can increase the value you derive from these assets.

Creating wireless communication links between mobile workers and physical assets can further reduce operating costs and improve your ability to respond to business or service needs. For example, a municipal utility's manager in the field is forwarded a report about a problem in a neighborhood. The manager uses a mobile device to access data being transmitted from equipment in the reported area to evaluate the problem, and then searches a Web-based map to find the nearest vehicle carrying the right people and resources to fix the problem. The manager sends them a dispatch order with directives for appropriate action.

To quantify the impact of municipal wireless in this area, measure the potential benefits of enabling mobile employees to access the Net and the city's network servers, to process information from anywhere at any time. Daniel Aghion, Executive Director of the Wireless Internet Institute reports that, "by all accounts, municipal and county governments in the United States employs upwards of 6 million professionals. About 2 million are mobile workers in a wide array of local government functions: law enforcement, emergency response, public works, social services, various types of surveying and inspections, utilities etc."

Much of a mobile city employee's work involves coming into the office to load up on paper-based forms and data, going into the field to perform services and gather data, also on paper, and returning to the office to type data on paper reports into a computer. Even workers who are radio dispatched, such as those in public safety or local utilities, often have their efficiency and productivity curbed by how much printed data they can carry, or how much data can be transmitted over cellular or radio networks.

Many mobile workers spend several hours a day searching on-line databases, filing reports and performing collaborative functions. Given today's deployed mobile communications options (essentially data over slow cellular voice networks or expensive proprietary radio networks), performing those tasks must wait until workers travel back to their headquarters. "This consumes anywhere from one-to-two hours a day of work time," continues Aghion. "At an average hourly employee cost of $25.00, enabling such tasks to be performed from the field could potentially take upwards of $10 billion of inefficiencies out of US local government operations. Similar savings could potentially be achieved in most of the developed world." What are these inefficiencies costing you?

Putting the numbers in play

Medford, OR, with a population of 65,000, deployed a MeshNetworks WiFi network from Motorola primarily to replace public-safety's CDPD network. At the same time, project leaders evaluated the potential benefit to all city departments to galvanize their business case.

Doug Townsend, Director of Technology Services for Medford, says "When we first began to build our business case, public works came back with initial numbers that made them the largest beneficiary from mobile technology. They calculated it would save them about an hour a day per crew. After we deployed, the second in command in the Building Department came up with a list with about 20 other benefits that we did not anticipate. Initially, they were one of the lowest in terms of an anticipated ROI , and now they are realizing significant advantages."

Taking a pen and paper to work through the following questions gives you a sense of the financial impact of wireless on the inefficiencies of a paper-driven operation within your organization. This example applies to any department that does inspections of one sort or another.

* How much does the average mobile inspector cost you per hour (salary, benefits, etc.)?

* How much does an office worker who processes inspection paperwork that's generated in the field cost per hour?

* On an average day, how much time does an inspector spend waiting for assignments at the beginning of the day, completing paperwork manually and traveling back and forth during the day to pick up new assignments?

* Multiply this time by inspectors' average hourly cost, and then multiply the total by the number of inspectors on staff. How much money is this?

* For a typical day, how many hours does an office worker spend processing forms from the field and contacting technicians to clarify illegible entries? Multiply these hours by the worker's cost per hour, and then by the number of workers entering forms. How much money is the government spending here?

Now, envision the impact of having municipal broadband in place and each inspector carrying a mobile device.

All of the day's assignments and any updates during the day are wirelessly delivered to inspectors' devices. Travel and morning wait time are eliminated. With drop-down menus and check boxes prompting workers through their tasks, inspections and data entry happen simultaneously, so that time spent completing paper forms is eliminated. Each inspection is faster and more accurate. Data from the devices can be uploaded directly to the city's computer network server wirelessly after each inspection. Subsequently all of the forms processing time by office staff is eliminated and there are few or no errors with which to contend.

Take a detailed look at all of your departments. As with the previous strategic areas, isolate activities and associated costs that can be impacted by field workers' wireless access. Also look at various city offices and facilities where in-building wireless access to e-mail and data instead of wire line access can cut costs.

Leave no stone unturned

As you do a citywide audit of your operations, you will likely find similar cost savings potential in all departments and among your partner organizations. Case workers from economic development groups visiting small business owners at their premises as well as public health and safety workers in the field can substitute digital reporting for paper processing. Street repair and maintenance employees can eliminate the paper processing from dispatch and job activities. In each case, you can cut costs and increase efficiency.

Don't forget to audit asset management and maintenance, which is a huge cost in larger cities. Office equipment, port-operations machinery, buildings the city owns and vehicles of all types. A city worker in Philadelphia found sitting forgotten in an old warehouse a major cache of artwork collectively worth $2 million that had been donated to city schools. Simple wireless sensors can keep track of items such as these. By how much can personnel increase the value and longevity of assets with mobile devices, wireless tracking and forms-processing applications to better schedule preventative maintenance or fix problems as soon as they occur? How much will efficiency improve if people can find things faster when they need it? If 100 workers spend 20 minutes at the beginning of every shift hunting for items misplaced by the last shift, the total annual productivity loss is high.

Also consider how much cost you can eliminate by un-wiring many of the city and county office buildings and other facilities. Can you eliminate T1 line charges with WiFi networks? Can you avoid a lot of re-wiring costs as you expand or switch offices? Maybe you can save lots of hassles during emergencies and natural disasters by using WiFi as well as fiber to back up communication systems within these facilities.

There's more than money involved here

By how much will you improve the level of service when inspectors start making better, faster decisions in the field because they can access the entire history of a building and its owners' transactions with the city while conducting an investigation? How much safer will your police or firefighters be if they can access that same information, along with an inventory of hazardous materials and images of floor plans before entering a building to respond to an emergency situation?

Look at how the immediate access to infinite amounts of data and online resources can improve the speed and raise the level of services provided by all departments whose employees deal directly with constituents at their home or business. Workers can better answer questions, have clients complete applications and other paperwork, and point citizens online to the most appropriate solutions for their concerns. Overall, you have happier citizens when city workers are meeting their needs quickly, with little hassle and greater competency because workers have more knowledge at their fingertips.

Affordable broadband access gives city governments as well as support organizations the ability to deliver new services. School departments can develop new Web-based educational programs that increase parental involvement with their children's homework. Job assistance groups can create mobile task forces that perform previously impossible same-day job placement services right in the neighborhoods. Tourism workers can capture real time feedback and pictures from tourist for news portals that are much more effective than passive, static infomercials on hotels' in-room TVs.

Another key element of your audit should be a review of joint operations carried out with civic, business and neighborhood groups, county, state and federal agencies, contractors and other entities that enable you to deliver services. A dramatic example is the coordinated response to major disasters.

Local public safety personnel, state branches of the National Guard, non-profit groups, insurance claim agents and others should woven together in a mutual drive to bring relief to the area. But are they linked sufficiently in a communication network? Can broadband wireless be the superglue that holds these often disparate entities together in times of crisis and raise every group's performance level?

Even in the most mundane aspects of everyday business, city governments rely on various third parties to support its service delivery and administrative operations. Many of these partners have mobile workers or offices that can benefit from municipal broadband. Mobile workers for charity groups who are delivering food and other assistance to seniors or low income communities can be more efficient by tapping into city data. Will construction contractors for your city projects speed up job completion and minimize disruptions if they can easily tap into building records, maps and appropriate city forms?

Communicating more effectively to prospective constituents

With commercial business organizations, this strategic aspect of wireless deployments typically refers to direct communication to people who potentially can buy what they're selling. However, for the discussion on municipal broadband, it really is a discussion about how you can use the fact of your technology deployment to draw people to your community.

Bragging rights for being wirelessly connected will probably have the least directly quantifiable benefits for many municipalities. However, city government, tourism boards, convention centers and chambers of commerce find that having their entire cities wireless enabled is a notable factor in individuals' and businesses' decision to patronize their town or city. If you don't have it, a lot of people may decide to go elsewhere if they have an "either-or" choice to visit you or another municipality that does have wireless. More likely than not, wireless access will become one of several key factors that, when taken together, convinces people to visit or move into an area.

If theaudit of theprevious strategic operations proves the economic or public policy value of broadband wireless, this fourth strategic area is the icing on the cake. "We can save a lot of money in our communications with constituents, and oh, by the way, we will make our city more competitive when bidding for more convention business."

Once you make the decision to pursue broadband wireless, be ready to heavily promote the fact that you have it. Within the next couple of years, having municipal broadband will be an expected item on the tick list of "Reasons to Move to Our Fair City," right along with good schools, favorable business ordinances and access to great transportation. In fact, an article in the September, 2005 issue of National Geographic Traveler magazine named Philadelphia America's "Next Great City" and cited WiFi as one reason. The article credits Philly's wireless plan for the recognition. "How do you knit together a rapidly diversifying city?" the magazine asks. "Use invisible thread."

Take a good look at what's going on now to promote you city. Will city- or region-wide wireless access increase your ability to bring new constituents to town? Is not having it likely to force your business to other locations? If the answer to either is yes, then you need to factor this into your decision to deploy the technology. Remember the story in Chapter 1 about Scottsburg. It's a stark example of the "play big or lose big" decisions that smaller communities face in whether to deploy wireless to keep companies in town.

In all four of the strategic areas, these audits will unleash a flood of ideas and suggestions, so have departments and groups use some sort of screening and prioritization process for these. You want to have a big picture view of the many possibilities and the potential financial impact and other benefits so you can make the business case for citywide broadband. But if you try to execute on too many ideas at once, you may not get anything started or you will overwhelm the people trying to implement these projects and many will never get finished.

The next section presents categories of tactical activities that you can use to help achieve the goals you establish. This is also a good structure for categorizing the ideas that are generated during your audits. From here they can be refined into an action plan that includes timelines, resources allocated and all of the other components of a good project plan.

Good tactics are the cornerstone of successful strategies

With any good strategy plan, there has to be a collection of tactics that enable you to make the strategy successful. This section contains an overview of six categories of tactics that collectively help municipal broadband and applications generate high dividends for your investment. While you're reading the book and ideas come to you, place them in one of these categories.

When you finish reading and begin directing the group of people working on building the business case for wireless, systematically review these tactical options to generate additional ideas for specific projects. The group can filter, organize and prioritize these ideas, plus those coming from the city departments, for your final technology implementation action plan.

1. **Eliminating the paper trial**. The amount of paper forms, reports, memos, etc. that is clogging organizations' business process arteries is staggering and incredibly expensive. The cost of processing paperwork from the field, constituents and various third parties is equally high, and a major drag on organizations' efficiency, productivity and cash flow. Mobile and wireless technology can generate great ROI just in this tactical area alone. Tactical efforts in this category can be applied to the first three strategic goals.

Providing workers with mobile devices and forms processing applications is one tactic, but can you add bar code readers to devices to scan physical assets? Switching workers from basic cell phones to cell phones with text messaging capabilities is a simple tactics to eliminate paper-based dispatching processes for vehicle drivers. A print or direct mail campaign is an example of a tactic to move businesses and residents from ordering paper forms to filing online and receiving text updates on their cell phones changes in filing procedures.

2. **Increasing brand awareness and loyalty**. In government terms, this means capitalizing on city-wide wireless to promote and reinforce the positive image of

your city or town among both residents and visitors. There were several pilot projects in Philly that, by providing wireless access to content that actively involved people with events in the city, gave these individuals a sense of greater connection with, and subtly increased their pride in, the city. Though mostly used in a business context, loyalty to a brand – in this case, a city – is what keeps talented college graduates, professionals and a lot of the residential tax base from leaving.

Traditional publicity campaigns and promotional offers to get key constituencies to try city and community portals are an effective tactic to get them using the network. Adding features that enable and encourage people to contact their friends or out-of-town relatives and invite them to the portal is a good way to increase word-of-mouth awareness.

3. **Generating immediate responses.** These tactics are effective ways to get people to take immediate action in response to communication from city departments and allied organizations. Once you have municipal broadband, workers equipped with wireless-enabled PDAs or laptops and constituents with wireless access on their person and within their homes, your ability to generate immediate response skyrockets. You can get field workers to go immediately to where they're needed, while you can get instant updates and progress reports. You can get the citizenry to respond faster to emergencies, deadlines for receiving services or to general events occurring within the city. And of course, constituents can use the technology to get a response from government that is quick, appropriate and efficient.

Generally, the core set of immediate response tactics is going to require a system of collecting and centralizing data by the city or its partners, and applications that allows pertinent information to be manually or automatically pushed out to constituents. You need to have some type of opt in procedures to ensure everyone's privacy. Text messaging is probably the most reliable format to support these tactics since they require very little bandwidth and manage to get through even when natural disasters knocks out most forms of power and communication.

4. **Educating your audiences.** Educating your different audiences about the various aspects of your city's operations, services and opportunities is a valuable practice that you can significantly enhance through broadband wireless. But it's not just about you educating constituents. There's a lot that people need and want to learn for themselves through traditional education, on-going skills enhancement and general knowledge gathering which wireless access facilitates. Delivering and encouraging these capabilities can be a significant way to attain all of your strategic goals.

You have many potential tactical options here, which increase dramatically when you bring in commercial and nonprofit partners. Self-help options are generally better since many people learn best at their own pace, and constituents have the benefit of 24/7 access. It's also good to have programs in place that enable constituents to find out things from each other. People in their teens and 20's thrive on mobile devices to stay connected with friends. Leverage this trend with your tactics.

5. **Demonstrating services.** City governments and community groups that are trying to increase the number of underserved constituents accessing services should consider inexpensive wireless access as a channel to allow these citizens to "sample" their local government. Citizens often drop out of the system because they feel it doesn't work for them. Your plan for deploying wireless should include access to online tools or content to enable residents to receive some type of useful service. A local "resource locator" feature, a notification system that sends alerts when a particular type of job ad is posted, a transit planner that enables people to find the faster bus and subway route(s) to city locations are all effective tactics. Your activities in this category likely require the support of several technologies and partner organizations.

6. **Executing more effective research for less money.** Wireless access enables two-way dialog and real time qualitative research by giving those with whom you do business quick access to Web-based and e-mail-delivered surveys, instant messaging and online chat areas. Besides wireless access from homes, wireless kiosks in various parts of the city gives citizens many opportunities to give you feedback so you keep your finger on the pulse of the needs of various constituents.

Putting digital forms on PDAs to facilitate traditional data collection by mobile workers for activities such as medical research, census taking, transit rider studies and "citizen-in-the-street" interviews is an easy way to save lots of money. The next chapter presents a good story about a Canadian city that used this tactic effectively.

To sum it up

Turn your city department managers, partner groups and other stakeholders loose with calculators to do a top-down analysis of their respective business operations. Look at those business processes that wireless can enhance or eliminate in order to save money, make money or run a better government or constituent service organization. Your three most frequent questions to ask are 1) how much does it cost in time or money, 2) why do we do it this way, and 3) what if we use wireless to....

Follow the framework presented here to organize everyone's thinking, areas of focus and ideas that are produced. Use the tactical categories to further organize ideas that are generated, and the actions that people plan to take as you pursue municipal WiFi deployment. Adapt as necessary to everyone's unique needs.

As you work through your audit, consider the suggestion from Intel's Paul Butcher that you "create some type of verifiable assumptions, and use the city's financial records to support these assumptions making sure you standardize benchmarks and values. Then have an accounting firm come in and verify to the tax payers the validity of your numbers. This would be an exciting way to let constituents know that you've done your homework well and there is sound fiscal basis for your pursuit of broadband wireless."

To give you an idea of the line items your assessment might produce, Jonathan Baltuch, president of MRI which develops and implements economic development solutions for local governments, has some hypothetical numbers. He helped drive the business model in St. Cloud, Florida, which offers broadband wireless to every resident and business for free based on the economic return of using the network to reduce the city's overall operating costs.

"School systems typically have a T-1 to service each school. These typically cost between $600 and $1,000 per month per school. With a municipal wireless system in place, these can be eliminated and the schools can receive 10 times the T-1 connectivity speed and capacity. That's a $120,000 annual savings if you have ten schools. Most cities today utilize laptops with cellular modems. Typically, monthly costs run from $40 - $60 per car for slow data transmission. If your city has 100 patrol cars, that's $48,000 - $72,000 per year that municipal wireless can eliminate. A municipality that realizes a productivity gain of 3% for 600 city employees by replacing paper-based processes with wireless applications could reduce the need to hire 18 employees at an average cost per employee, including benefits, of around $50,000 which is $900,000 annually."

You now have a high-level view of your strategic and tactical options and thumbnail sketch of the potential benefits for using broadband wireless to reduce costs and improve operations at all levels of city and county governments. Add to these and fill in the details for your organization as you read the following chapters that tackle the logistical issues Philadelphia addressed. To get a comprehensive presentation of these points and examples of commercial, non-profit and government organizations' wireless projects, get a copy of my book "I Only Have ROIs for You," published by Hudson House Publishing (www.successful.com/booksummary.html).

The next chapter presents a variety of stories on city- and county-level implementations of wireless applications. These snapshots of municipalities, their partners and stakeholders show you how they are making wireless work for them and provide more fuel for your idea generators.

Chapter 3

Oh Brave New Wireless World: Municipalities embracing change

Now that you have a framework for auditing your business operations to determine where and how wireless makes sense, let's move from the theoretical to the real world of municipal governments and their adoption of wireless. This chapter presents a series of stories of how governments, as well as organizations that complement government work, are using wireless. Use these to further develop the vision for your implementation plan.

Cities on the move

Citywide deployment of wireless technology is in the embryonic stage in many locations, but wireless applications have been in place in quite a few cities for several years. Most of the city departments highlighted here, including License & Inspection in Philadelphia, are actually pioneers given that they started wireless initiatives before municipal WiFi became a hot topic. But plans are definitely in motion to integrate some of these efforts into the bigger picture of broadband everywhere.

Philadelphia's License & Inspections Department leads a new-age revolution

This department's wireless efforts are actually part of an important first phase in a process that will transform how city business gets done in Philadelphia. Wireless is one component of a new automation system that will integrate the operations of various city departments using both desktop computers and mobile devices.

The Department of License & Inspection (L & I) enforces Philadelphia's building, fire, housing and business codes, issues licenses and conducts inspections. Like many departments and agencies in large cities, L & I's field staff is perpetually challenged to keep up with the service demands of numerous citizens and businesses. "So there was a prioritization and, similar to the squeaky wheel, the more prominent things received attention at an adequately acceptable level," states Computer Information Systems Manager James Weiss. "This new automation system allows us to keep track of all the demands that are out there."

L & I is in the final stages of customizing commercial off-the-shelf code enforcement software from Hansen Information Technologies. The wireless component enables building inspectors with mobile devices to receive information and submit paperwork while in the field. According to Acting L & I Commissioner Robert Solvibile, "the goal is to let them work from the field. One reason they come into the office currently is to hand in paperwork, and the other is so we can monitor their progress. With the system, we'll be able to do both while they stay out doing their job. Maybe once a week they'll come in for training and to get department updates."

From analysis during the pilot project, the department is convinced that wireless is going to save a lot of paperwork processing time. A weekend on-call supervisor can receive as many as 20 calls about code violations. Without wireless, this person has to complete the forms by hand, give them to their supervisor to screen, and then take the forms to the clerical pool and have them print it out and send notice to the owner. It takes four days at best. But with this system, as soon as inspectors input data to the devices it can be sent to the server. A batch run happens that evening and the next morning the printed violations go out to building owners.

"As far as the money value of it, I can't tell you at this time what that's going to be, says Solvibile. "We don't know how much savings we'll get, but there will be a great savings of inspectors' time in the field. And with the workload we have, no one will be losing their jobs over this, they'll just be getting more done in the same amount of time. More importantly, if you're a citizen who has a concern about a violation, it's important to get a notice into the building owner's hands immediately."

Not only is L & I happy with the paperwork reduction potential, but also the increased efficiency in how inspectors do their jobs. One problem with paper forms is that, if an inspector writes and makes a mistake which gets typed up in the office, there is a ripple effect of wasted time when someone in the office or another department has to reference that building's file later.

"In this case, we have pre-populated drop-down boxes and the whole bit," says Solvibile. "Inspections are faster and violations are printed the same way each time and in terms the recipient will understand. 'Code section such and such says you must have the following equipment in your bathroom.' This could be written a million different ways by our inspectors in the paper system - a toilet, a commode, a wash basin, a sink. Now everything will be the same every time, and you won't have those handwriting mistakes."

More than paperwork reduction

Besides the efficiency improvements that come from digitizing paper-driven systems, the application's value is extended through its ability to integrate data from different applications and also enable workers in different departments to share data. To get to this point required several iterations of software.

"The first product that we got didn't give us the integration abilities that we wanted," states Solvibile. The inspectors could process violations, but they couldn't see the history of the buildings and any transactions the owners had with the City. "Just a few weeks ago we got the component that will give us this capability. Inspectors will not only know if any violations are outstanding, but also all the licenses and permits that a property has received since it was built." It also enables the complaint processing software to put data into the wireless application so inspectors can see any complaints made in addition to the violation. Another form doesn't have to be processed and handled by the inspector.

L&I plays a key role in public safety, and the system's ability to share data is a big plus for the city. They enforce the fire code and the Fire Department puts out fires. You come in today and get your building plans for a construction site. L&I approves them and then these plans are imaged and stored online. If there's a fire, fire fighters en route will be able to effortlessly bring up these plans on a mobile device to see what they're up against, what's there, how the place was built and so on. It's going to be invaluable.

As citywide broadband comes online, L & I sees another wave of improvements to their business operations resulting from the shift from the cellular network they're currently using. Weiss reports, "with the CDPD network we found that coverage as a whole wasn't ubiquitous. There were many areas where, regardless of the time day or night, you couldn't get a signal. Inspectors could work on devices even when they didn't have access. But clearly the optimum situation is having the technology enable inspectors to quickly get the most up-to-date information into the main database so everyone who needs to can act on it. The Wireless Philadelphia project moves us towards a ubiquitous and less expensive network. It should give us more throughput."

Weiss also believes that the city as a whole benefits. "The notion of broadband access available to 100% of the citizens is something that's going to allow us to extend services. We'll see the migration of more of our customer requests for information move from in-person visits or telephone calls to Internet transactions." The answers won't be hand-me-down from a neighbor who used to know somebody who worked for the Department. You can find out from the same source of requirements, spelled out in plain English, that the employees look through. Everybody will be looking at the same rule book, in effect.

More citizens will be able to do a significant amount of license transactions online through an instantaneously available, convenient process which starts with the basic question of "Do I need a license for this?" This ability to step off on the right foot with the right regulatory entity in a way that's convenient and economical is going to make the process better for everyone. L&I won't be sending constituents needlessly off to different places. They'll deal with licensing in an intelligent way and faster."

Oh, Canada! Mobile tap dancers count trees, save the day

One city. 52,000 trees on 45,000 properties. Four student data collectors to count 'em in four months. Each and every tree. Is this one province's Waterloo? No way! Mobile technology saves the day - and over $100,000. It's a great example of the tactical option described in the previous chapter, executing more effective research for less money.

The city of St. Catharines in Ontario, Canada was updating its data base of trees that are growing on its residential, commercial, industrial and parkland properties. "We were trying to follow a five-year maintenance cycle to recount the trees, measure them, and record their species and location," said Joe Keri, Systems Specialist. With so much data collection, using paper forms or hauling around heavy laptops was out of the question.

Keri opted for mobile devices as the solution, and used AppForge's Crossfire development tools to custom-design an application in three weeks. With 10 screens of drop-down checklists on HP iPAQs, the mobile fab four took to the streets "tap dancing" through the screens to not only count every tree, but also to record soil conditions, landscape features, tree health, site encumbrances and abatements. They even included information on city properties that didn't have trees to track them as potential sites for adding new trees.

The students - "they weren't even horticultural students, they were business communications majors" - captured data for hundreds of properties daily, and used desktop syncing to dump the date into Oracle databases.

"I've been doing this kind of work since 1976," states Keri, "and I've never seen a job of this size done so quickly and efficiently. Before we decided to use mobile devices, we weren't even sure we could do the project. At least not in the timeframe we had." It would have taken a whole army of data collectors using paper.

Keri estimated it would have cost about $125,000 using conventional methods rather than the $20,000 it eventually cost using mobile technology. He now has plans on the drawing board for new data collection projects. Things are looking rosy in the Garden City.

In Corpus Christi, TX a single seed bearing multiple fruits

Similar to Houston, Corpus Christi, Texas, began their muni WiFi effort with a single focus on an application based on ROI calculations that justify the network's cost, and now the city is seeing potential benefits in other areas. The city's $7-million, 2,000-cell Tropos WiFi mesh network enables automated meter reading (AMR) of city-owned gas and water utility meters. The meters transmit data to a central server, allowing customers to keep tabs on their daily usage online. The system cuts downs on mis-readings as well as mishaps to meter readers accessing difficult properties. Close monitoring helps utility managers match gas usage with gas price fluctuations and control water flow to reduce system breaks.

But AMR is just the beginning for the city. Pronto Networks' software-management platform provides IP-based security, the ability to manage multiple subscription services, and enabling public access to government and schools Web sites. Some students in Corpus Christi are beginning to receive WiFi enabled laptops to facilitate their access to the network.

"When we started looking at the bandwidth, there was an awful lot left over to do some things for folks who want to be roaming about wirelessly in the city," says Leonard Scott, MIS unit manager and project leader. Of the city's 3,000 employees, the 70 percent who work in the field will benefit from dozens of planned mobile solutions. For example, in the event of a bank robbery or remote tactical situation, public-safety responders will click on links that take them instantly to security cameras. The technology can be expanded to container operations at the city's large port and will accommodate smart-chip technology for monitoring firefighters entering burning structures.

The city is also working with hospitals to provide citizens with their medical history embedded on chips that are enclosed in medic alert-types of necklaces and wrist bracelets. These can be read by emergency responders in the field who can immediately pull information about allergies, medications being used, previous surgeries and so forth, then immediately dispatch this to hospitals.[1]

Who let the dogs out! Woof!

Lincoln, NE's city government is committed to serving constituents by making theirs a 24/7 wireless-enabled organization. The first department to deliver this level of service is the Lincoln Animal Control Division.

Using wireless PalmOne devices, Animal Control's staff access their mainframe computer to get data on animal owners and pets, and review previous dispatches from virtually anywhere within their jurisdiction. In its first year this implementation decreased the cost of distributing information to the staff by 50%. They also eliminated the need for monthly printed lists of pet license numbers that quickly would become outdated. Now, when an animal control officer finds a stray dog or cat, locating the owner is as simple as entering the license number on the animal's tag on the PDA. The officers also get background information about the owners as well as history on the pet.

The increased efficiency witnessed in Animal Control is catching on, and other city departments are just going to the dogs. The Property Assessment office is using PDAs to get real-time access to current assessment records and updating information from the field. Weed control inspectors are using PDAs to track violations. The city's next project is working on a criminal justice application that searches databases for license plate numbers, VINs, and outstanding warrants.

---

[1] Source: Intel white paper "Digital Community BestPractices"

New York, New York loves wireless

New York City's Department of Health and Mental Hygiene and the Bureau of Food, Safety and Community Sanitation has a small army of public health inspectors in the field every day armed with mobile devices and one-pound battery-powered printers. Among other things they inspect the Big Apple's better (and better to not know about) restaurants. In some cases they issue notices of violations which requests the honor of the owner's appearance in court. These are printed in the field and the data is transmitted to an administrative tribunal.

The inspectors use the devices to handle all of their inspections. Rather than pop up a rote set of questions for inspectors to complete, the software is designed to prompt questions based on the situation at hand. Small restaurants have fewer questions than larger ones. If a violation is noted, different questions are queued up. Time, date and key information about restaurant owners are automatically stamped on the report so there's no room for errant or faked reporting.

Prior to deploying the mobile devices, inspectors used paper forms that resulted in duplicated data entry, which was time-consuming. If certain information on the printed forms was wrong, a violation could be thrown out of court. Handwriting errors wasted judges' time on deciphering through the scrawl, and wading through cabinets of previous paperwork wasted inspectors' time.

The mobile devices and well designed software saves time, increases the swift administration of justice, judges are happier, the home office has reduced file storage space and inspectors are covering more territory than before. Overall, the public is better protected and served. There were some issues that had to be worked out during the pilot project, but once they went to full deployment the department realized ROI within a year. Now that the city is moving forward with its exploration of broadband wireless, this and other city departments are likely to become more aggressive in their wireless initiatives. Bon appetite!

The fire burns hot for wireless in San Diego

SD Medical Services' mobile application was a lone-wolf effort that started because the department Fire Chief and the Electronic Documentation Coordinator were frustrated with the process for recording and tracking patient services. Particularly limiting were the data input options of the printed forms. Personnel in the field often didn't follow the rules when completing forms and their handwriting was often illegible.

Chief Greg George and EDC John Pringle developed the prototype in-house for a mobile application that runs on palmOne devices and uses Intellisync's Data Sync to move data between the handhelds and servers, and took it to management. "We are both City of San Diego employees who work for the fire department," states Pringle, "but the fire department is partnered with Rural Metro Corp. a private company, to provide patient transport in the city. Without private participation, this project wouldn't have happened."

The pilot ran for six months with 70 people. Then the application was rolled in an initial phase to everyone (150 people) who used it for 16 months. Based on feedback during this period, San Diego Medical Services rolled out a phase 2 deployment in which software, server and devices were upgraded.

There was no formula in the beginning, so everyone had to rely on anecdotal information to gage how successful the application was. "After everything was in place we started to see things that the system enabled us to measure. We had billing records so we used these when looking at data from the new application, then started making comparisons."

For years, the department was losing anywhere from four to 15 patient documents every day (200 patients are serviced citywide daily). It would be two months before they discovered which reports were lost, so getting data at that point was impossible. With the mobile application there are zero records un-accounted for, eliminating a $30,000/month shortfall that had resulted from not being able to bill for the patients whose records they didn't have.

The palmOne devices can capture signatures, plus every day workers have to reconcile their patient records with a master digital printout. So now, on a Monday the crews do their runs, by Tuesday all of their data is sent electronically to the home office and on Wednesday the home office bills for these services. It used to take five days getting from service delivery to service billing.

City and county officials like the system because it prompts workers with the right procedures (medicine dosage, for example), and the screens changed based on type of patient, offering the appropriate questions medical workers need answered. The system produces more details than paper forms were capable of capturing and subsequently raises the bar for worker performance.

The liability factor played a role in determining ROI because patient records sometime end up in court. "Having the system force people to follow all the proper steps plus generate extensive electronic data ensures that our documentation is done properly, legibly and completely," reports George. "This protects the City, and our employees from liability."

City stakeholders adopting wireless

There are a number of organizations within a city that, even though separate from the government structure, have an equal obligation to operate at maximum efficiency in the interests of the various constituencies. This section takes a look at several whose operations have a big impact on their respective cities.

Come fly with me to reduced costs, greater ROI

Airports are vital to the transportation in and out of municipalities of all sizes, so there's very little doubt about the symbiotic relationship here. FMC Airport Systems, sells airports cargo loaders, passenger bridges and ground support equipment. Don Pohly, Business Manager for FMC developed ROI numbers for using embedded wireless technology to reduce equipment requirements, increase employee productivity, reduce aircraft damage incidents and reduce flight delays at airports. This story is important because aside from showing how to make airports more efficient, you get a good feel for how embedded wireless can have a major financial impact on managing your city or county's valuable assets.

At one major hub, an airline has ground support equipment (GSE) with a total value of over $30 million. Management's goal was to reduce the total quantity of equipment by 5% through the use of real time location tracking and telemetry (measurement of the performance of mechanical equipment) technologies. The eventual total savings was closer to 7% which represents $1.5 million in capital expense, about $300,000 in annual depreciation, and about $225,000 more in annual maintenance costs.

An estimated 1/2 hour is lost at the beginning of each shift as lead employees search for the equipment needed for their crews. Another 1/2 hour is lost per day in certain paperwork that could be automated. Using a wireless location system to track the equipment and a telemetry system to complete paperwork wirelessly saves an hour per day for up to 60 lead employees, or over $500,000 per year in increased productivity.

In addition, employees looking for items throughout the day waste a tremendous amount of time: aircraft-handling staff, baggage people searching for specific carts, cargo people looking for the right containers, maintenance staff hunting for equipment that's due for maintenance. The value of reducing this wasted time is difficult to estimate precisely, but conservative analysis put it at over $100,000 annually per airport.

Aircraft damaged due to GSE mechanical failure might happen just twice a year at a large hub airport. But even a relatively minor aircraft repair can easily top $200,000. Add to that the direct expense of a cancelled flight and the indirect cost of customer dissatisfaction. Preventing even half of these incidents produces significant savings. Having the right equipment well maintained and ready for use in the right place, at the right time also helps cut delays in flight departures. The cost of *each minute* an aircraft is delayed has been valued at $55, so saving only 40 minutes per week in delays nets over $100,000 in annual savings.

These savings combined with many smaller savings gained elsewhere in an airport's operation have consistently achieved a payback on an embedded wireless system in one year or less. This ROI has been reviewed and estimated with several airlines. Once a wireless infrastructure is installed, people find additional applications for the data gathered and this further enhances ROI.

If an organization on its own purchases an asset management system that uses WiFi as the backbone for making everything work can cost hundreds of thousand dollars for tracking devices, software, training, etc. But an airport working with a city from the initial planning stages should be able streamline these costs by relying on the city's WiFI infrastructure. Systems relying on RFID (radio frequency identification) or other proprietary technology can cost millions, and will only do asset management while WiFi can serve other business purposes.

37 tons and what do you get? Another day older and a need for wireless

Municipal utility companies' operational role within a city is beyond critical. Providing water or gas services and all of the attendant customer support functions are the foundation to the economic, social and even physical health of every citizen. Though it's a regional rather than municipality, this Northeast Utilities story represents just one way a utility can use mobile and wireless technology as the perfect cost exterminator.

Before going wireless, field service people for Northeast Utilities (supplying electricity in Connecticut, Massachusetts and New Hampshire) used to haul around 12,000 printed maps that pinpointed the locations of buried electrical lines and equipment. The costs? The price of 37 tons of paper. Since these maps had to be frequently updated, there was an additional annual cost for 1,800 computer printer cartridges. Adding to the expense were vans that had to be customized to increase storage space for the maps, plus equipped with special heavy-duty springs to carry the extra weight.

The ROI for laptops with wireless modems to store maps and receive updates while in the field? For a $40,000 investment, Northeast Utilities saved $383,000. It eliminated the bulk of the paper, the cartridges and the need for the specialized vans. Field service people are more productive because they access maps instantly rather than having to paw through piles of paper maps, plus updates are immediate which guarantees that the staff will find the problem the first time out. The staff is also more efficient because they can go where they are needed rather than be restricted to areas covered by the limited number of printed maps each van could carry.

Your paper costs in field service may not be as dramatic as Northeast Utility's. But there are other benefits such as a greatly streamlined dispatch operation. Within an area covered by municipal broadband, Northeast or any utility company can jettison expensive proprietary radio systems that can garble transmission and not produce an adequate or accurate paper trail. More importantly, people in the field can quickly receive real-time updates to graphic files of maps, building floor plans, pictures and diagrams of equipment to be serviced and so forth. IP-based cameras carried by repair teams can beam back live video feeds from work areas to get additional support from the home office staff who see everything people in the field are seeing.

A port in the storm builds brand loyalty with kids

The ability of a city's museums, historical sites and other destinations of interest to creatively attract and engage tourists is important to the overall economic development of a municipality and its businesses. Port Discovery in Baltimore, MD is a museum of exhibits focused primarily to children, and they added some very interesting wireless capabilities that have made them a big draw. Bringing your city's attractions into the muni WiFi planning process can enable some great ideas to take root and grow into wonderful crowd-building tactics.

With funding from the State of Maryland, the W.K. Kellogg Foundation and Aether Systems, Inc., Port Discovery opened a new exhibit, the PD Kid Club in July 2001. The PD Kid Club is Port Discovery's sixth and final permanent exhibit and culminates the museum's initiative to integrate community and technology.

Port Discovery's staff believes the digital revolution is changing the landscape of opportunity for kids, allowing them to express themselves, interact, create, and communicate in powerful ways. PD Kid Club membership introduces youngsters to several digital technologies that give them the experience they can apply to their own pursuits in everyday life.

To show kids how technology can empower them, The PD Kids Club showcases high tech methods of creative expression. They can use computers to create their own digital music mix. Members also can use computer software to create elaborate animated pictures and digital movies for playback on a PC.

Wireless interaction is enabled through the Kid Club Communicator (KC^2), a modified RIM BlackBerry that allows members to interact on a deeper level with the exhibits at Port Discovery. Kids use KC^{2}s to access more interpretive content in all of the museum's exhibits, interact with the physical environment in new ways, and keep track of the accomplishments they make in the exhibits.

According to Port Discovery's Marketing and PR Manager Michelle Stetz, the KC^{2}s are a fabulous success. "The kids and parents are very excited using the KC^{2}s because they're unique. We do exit surveys and find that adults like them as much as the kids do. One reason is that the devices get parents working together with their children rather than just sitting down while the kids go off and explore which was our goal.

Stetz says the museum benefits because they're now widely recognized for bringing new technology into this environment and using it in creative ways. The wireless capabilities are also valuable because the museum can significantly change all of their exhibits just by changing content. They avoid the expensive and time-consuming task of dismantling and replacing the physical exhibits, plus they can promote new programs more frequently, which increases visitor traffic.

The museum has a five-year plan to further enhance their use of Web and wireless technology. "We're a prototype for other children's museums that want to implement similar programs. We intend to expand how we use wireless capabilities in other exhibits, and we also want to increase the capabilities of our Web site so kids can use it interact with exhibits and activities at other museums. Who knows what the future holds? Someday wireless capabilities may enable our kids to interact with kids in other states." By having a communities of kids and schools linked into muni WiFi networks, museums such as this can expand their "in-house" applications to enable and coordinate citywide educational programs.

Nothing could be finer than wireless in Carolina

One of the more ironic aspects of the discussion about whether or not tax dollars should cover the $15 million to bring wireless access to all of Philadelphia's citizens when the city spends $150 million to build two new stadiums for its sports teams. Just understanding how the Carolina Hurricanes hockey team used wireless to build game attendance should give some cities creative ideas on how to roll some of the funding for muni WiFi into their local sports organizations' marketing budgets.

The 'Canes in Raleigh, NC have jumped on wireless communication with both skates. They ran three wireless campaigns during one season to reach both fans and people who had never seen a game in their lives. The Canes Howard Sadel saw wireless as a huge opportunity to communicate with people in a different way and give them a new point of purchase. "A lot of people miss our games because they don't know about when the games are, or they get the information at time when it's not convenient to call.

"We did research to find out what was the best opportunity to push out ads since we knew we could target ads to specific geographical areas. In focus groups our fans said that they don't make entertainment plans any further out than three days because of schedule uncertainties, family events or other issues. We saw wireless as the best way to get something to them."

The first campaign consisted of ad banners on various wireless Web sites accessed by people's cell phones that just said "Purchase tickets for upcoming Canes game." The Hurricanes estimated that 15% of people who saw the ad called the ticket office, but they didn't have good tracking mechanisms in place to accurately measure sales results. They compensated for this with the second effort.

The next campaign was an ad with a promotion code so ticket sellers could better track sales. For three Saturday's the Hurricanes offered a $5 discount if the ad recipient called as they were looking at the ad. Callers got right through to a salesperson on an 800 number and could place an order. Again, 15% of those who saw the ad called and one third of them (5% of the people who saw the ad) actually bought a ticket.

The third wireless campaign was part of a bigger marketing effort designed to double season ticket holders from 6000 to 12,000 in an 8- to 10-week period. Sadel states, "We used everything: TV, print, radio, the Web. We had a Friends of Canes fan club in the corporate community promoting season tickets, and a local radio DJ broadcasting from the roof of our office building.

"On last eight days of the campaign we used wireless advertising. The phones became swamped during this last week and made it impossible to track leads to specific media sources. We do know that 7-game season ticket packages skyrocketed. We had 35,000 wireless ad impressions within a 90-mile radius with a 6.5% call-through rate. There were too many calls received to sort out conversion rates."

With citywide broadband in place, and given the rabid devotion of some sports fans, a sports team could unleash a far-ranging variety of campaigns to drive sales and city loyalty through the roof. Taken a step further, these organizations could band with local businesses to create a surge in local and visitor commerce that would make a chamber of commerce president absolutely misty eyed.

City partners take to wireless

In Philadelphia over 300 social service agencies contract with the city to provide much needed services to people in their respective communities. There are numerous similar groups providing services and support to business, education and health care constituents. Any medium size or large municipality no doubt has a comparable number of partners. One of the great values these organizations provide is the ability to tailor programs to meet the specific needs of the neighborhood. The key with making municipal broadband integrate with those efforts is extensive pre-deployment planning. The following stories give you a small snapshot of the endless possibilities.

Success one Philly neighborhood at a time

The People's Emergency Center Community Development Corporation (PECCDC) offers a wide selection of technology programs which bridge the digital divide and promote technological literacy for residents in their West Philadelphia neighborhood. Combining access, services and technical support, PECCDC's efforts to make technology more accessible complements its main goal to comprehensively transform neighborhoods into Communities of Choice where people choose to live, work, shop and raise families.

In February 2003, PECCDC began the Digital Inclusion Program to increase the technological literacy of low income families and increase their access to relevant human service information and tools. PEC President Gloria Guard states that "we realized that if we can't get computers into the homes of people in our community, children in those homes will not be able to keep up with their classmates who do have computers. They won't even be able to apply to many colleges, so how can they compete effectively in the marketplace?

Once connected to the Internet, families use this resource to pursue educational opportunities, search for employment, identify work-readiness programs and help their children with homework. PECCDC partnered with the United Way of Southeastern Pennsylvania's Teaming for Technology (T4T), Philadelphia Neighborhood Development Collaborative (PNDC), Achieve Ability and One Economy Corporation (OEC) to make this project a success.

Over 135 people have participated in the program to build self-reliance and strengthen opportunities for advancement. Given the program's success, Wireless Philadelphia Executive Committee chose them for one of the initiative's pilot projects.

The Digital Inclusion Program uses a four-pronged strategy to advance technological literacy by providing 1) computer distribution and affordable wireless access, 2) training, 3) relevant and useful content and 4) technical support. PECCDC sells affordable refurbished computers for $120 and provides wireless access for $5/month for the trial/initial year by splitting its T1 line to provide neighborhood WiFi access.

Participants in the program learn how to operate computers, use email, navigate the internet and are introduced to basic software applications. They also use One Economy's portal called The Beehive (www.Beehive.org) which is a one-stop resource center that provides links to government social services and local community-based resources. PECCDC provides ongoing technical support to participants.

Closing the divide one citizen at a time

Winnie Harris is one neighborhood resident who, after being downsized from her job and facing serious unemployment prospects in a rough economy, joined the Digital Inclusion network users. To her, getting back online seemed as basic as getting electricity service. Harris has come a long way since initially using her computer to file unemployment claims. She has developed a business plan for her own interior design firm, which she has launched and manages using her many new computing skills.

As part of the Wireless Philadelphia initiative, the program continues to grow and expand services. The Ten Technology Network is a ten week after-school program that offers about 50 high school students the opportunity to advance their skills in the field of computer technology. Combined with a Kaplan reading and mathematics course, students are introduced to basic computers skills while also learning to refurbish computers, design websites and navigate the internet. Some of them continue working with PECCDC to provide training and tech support services to others in the Digital Inclusions program.

In cooperation with the Pennsylvania Department of Public Welfare (DPW), PEC pioneered a pilot e-government system that enables families who are eligible for public benefits to communicate, transact business, submit documents and resolve issues electronically with DPW over the Internet. This allows clients to meet the requirements of welfare legislation without disrupting their obligations to work or family through hours spent commuting to DPW offices and sitting in meetings. Dozens of residents have successfully participated in the E-Assist program.

Making sweet music with wireless in Motown

In an area best known for automobiles and soul music, wireless enables Gleaners Community Food Bank of Metro Detroit to help those on the down side of the American dream. For them, the technology is a great business operations tool that enables them to do more with the limited resources they have, and through that increased efficiency inspire additional sources of revenues to support the organization.

Gleaners serves five counties in this part of Michigan. They decided to install an inventory control/warehouse management system with vehicle-mounted terminals, handheld computers, bar code label printers and WiFi access points. Before doing this, the organization distributed 14 million pounds of food with six employees dedicated to the selection of orders. In the system's first year, Gleaners distributed 24 million pounds with three employees. All phases of the warehouse—receiving, put-away, replenishment and distribution—improved once the system was implemented. And there are tighter inventory controls and fewer mistakes in inventory management.

For Gleaners, mobile technology also produced operational changes that are related to donors and auditors. "One of the great things we do with our donors is to walk them through the warehouse with a handheld unit," says Agostino Fernandes, VP of Operations. "We find one of their pallets and scan the pallet tag. If the tag reads 30 cases on the pallet, we can count those 30 cases. It demonstrates that we are great stewards of their donations." Because it's a non-profit organization, Gleaners is subjected to myriad audits. Auditors have changed the way they work when they visit Gleaners by now confirming the inventory count using the handheld scanners.

Within cities with municipal broadband, an organization such as Gleaners can extend its operating efficiencies to its workforce in the field. By having all of its employees in the field linked through mobile devices, dispatch operations can be improved, deliveries speeded up and reporting made more effective. Organizations receiving assistance can easily tap into food banks' networks to improve the on-site operations where residents are served, as well as streamline ordering. Consider this the common citizen's supply chain management application.

State organizations can be partners too

There are numerous state agencies and organizations that contribute important elements to municipal operations, and local governments need to explore ways for municipal wireless initiatives to tap into these opportunities. The Washington Association of Sheriffs and Police Chiefs launched a statewide initiative that uses technology to resolve a major communications challenge to multi-jurisdictional law and emergency response efforts.

In the aftermath of the Columbine High School tragedy, a delegation of law officers went to Colorado to gather insights to improving emergency preparedness in Washington. One apparent issue was that, after arriving at the scene of a crisis, it can take first responders hours to gather the knowledge specific to the circumstance they face that is required for them to take effective action.

One of the county sheriff's offices created a database application to provide easy access to resources such as aerial photographs of buildings and surrounding areas, floor plans, internal and external evacuation routes, recommended staging areas, etc. This software was licensed to a software company called Prepared Response, refined and is now the heart of the Washington initiative.

Using the software and tactical response plans created through inter-departmental/constituent planning sessions, any number of fire, police and other first responder units can immediately pull up the same information on laptops. Maps point them to power, gas or cable shutoff points and lists of the right utility people to call for immediate action. Within minutes they can direct vehicles in and out of trouble spots to evacuate citizens, or have appropriate streets and freeways blocked.

Marrying the technology with municipal WiFi significantly raises the level of effectiveness because the massive databases driving this system can be instantly accessed and updated from the main network servers and from each other's laptops. Static floor plans and pictures can be augmented with live streaming video. Wireless instant messaging, text messaging and Web cams all become the universal communications link that resolves the problems of agencies' radio not being able to communicate with other agencies' radios.

How many Columbines, Hurricane Katrinas and 9/11's do we have to watch disaster compounded by disaster because agencies, departments and citizens aren't able to use the technologies readily at hand? State legislators, rather than passing laws to prohibit municipal WiFi, need to fund state initiatives such as Washington's to integrate local government programs and resources under wireless umbrellas that provide for the common good.

To sum it up

Municipal wireless is neither technology flight of fancy, nor lethal threat to the laws and order of the marketplace. Quite the contrary. This is real technology that is proven effective and reliable at improving the way organizations function as viable businesses delivering products and services to their respective customers and constituents. It's obvious from these stories that city governments along with their related partners and stakeholders have been using wireless successfully before muni WiFi hit the scene, and this latest iteration of the technology will extend and strengthen these pioneering projects.

These stories and the many more like them that are springing up worldwide are also testimony to the positive workings of market forces. The demands and needs of customers in the marketplace - city and county governments are just as much customers of technology as commercial businesses – have anointed wireless as a superior technology to meet many of those needs. As any savvy customer would, these governments have tasked sellers of the technology to address these needs in a way that is most creative and cost effective. Those vendors in the marketplace that can best fill the bill deserve governments' business.

The bottom line is that every city that hasn't explored the muni WiFi option needs to take a close look at this. "I firmly believe that every city needs a strategy, even if that strategy decision is we don't need wireless at this time," states Philly CIO Dianah Neff. "It's really important that they do the due diligence to determine what this technology can mean to them." Once you've completed you task audit, everyone's excitement and support for the vision should intensify to carry you to the next phase of implementation.

In the next chapter you will see how Philadelphia initiated the project team that gave wings to the Mayor Street's vision of "more people online, not in line."

Chapter 4

Philadelphia creates the winning project team

If you want to have a successful wireless project, assemble a great project team with all the right people. Of course, after stating the obvious, the obvious question is how do you do that? Governments generally don't have all of the same sticks and carrots commercial entities have to entice and encourage the best and the brightest.

Well, first, there is no universal "right" person or group of people with extensive experience in this area. But there are some common threads that will probably identify good prospective candidates. The internal and external politics of municipalities, combined with the evolving nature of muni WiFi, require that you assemble a team of individuals with a diversity of skills and social/professional backgrounds. So you're not really looking for Mr. or Ms. Right, you're looking for Mr. or Ms. Right for the Job at Hand.

Second, Wireless Philadelphia proved that you can take a group of volunteer business, education and social service professionals, team them with government professionals, mix in a few consultants and move mountains. The business plan for Philly's initiative, along with relevant research, and extensive needs analysis was developed and presented to the Mayor in three months by a group that had never worked together before. This rivals efforts by some companies with big bucks backing them up.

This chapter looks at the project team from several perspectives. There is the process of creating and deploying wireless broadband across an entire city or county. There are the efforts by city or county governments to wireless enable their internal business operations. And finally there is the effort to facilitate wireless applications and projects at the community level that are managed by teams of people representing the various constituent groups. Each of these major deployment activities may require its own project team, though some members of the team driving the citywide deployment are likely to have a role assisting the other teams as is the case in Philly.

Basic team composition

Philadelphia's project team that took the city from concept to the written business plan consisted initially of a leader (CIO Dianah Neff) who presented the idea to Mayor John Street and a staff person who helped with the first pilot. When the Mayor gave the go ahead for the project in the summer of 2004, an executive steering committee of 17 volunteers was appointed by Mayor Street. The Executive Committee began working on the plan at the beginning of September and presented the finished document to the Mayor in December. Varinia Robinson came on as a consultant in the Project Manager role and with several other consultants make up the current project team. The Executive Committee was disbanded shortly after the plan was completed.

Ms. Neff was and continues to be the main driving force and public face of the project, as well as ex officio member of the Board of Directors of Wireless Philadelphia (a position designated for the city's CIO). Ms. Robinson managed much of the daily technical operations for the project while coordinating the Executive Committee's activities and the community groups that launched the pilot projects. She has similar responsibilities now but deals with different individuals and vendors.

Wireless Philadelphia was established to select and manage the vendor that is in charge of the physical deployment, and to drive all of the digital divide and economic development programs related to the broadband network. The Board of Directors has nine members, four appointed by the Mayor; four elected by the existing Board and the *ex officio* Director who has full voting and other powers. The length of terms of the directors is varied, and their appointment dates are staggered to minimize disruptions due to changes in Mayors or the city political landscape in general. The Board will hire a CEO and all the necessary managers and staff.

For larger cities and counties, this approach to city and volunteer personnel arrangements may work well, particularly the non-profit corporation approach since it distances the city government from political and financial risk exposure. It also protects the project from many of the political machinations and enables it to run more efficiently as a business organization. As time goes on, other organizational structures might prove valuable as well. Cities that get early clearance and budget approval for the project might decide to set up a nonprofit immediately to let it create all of the necessary staffing structure.

Smaller cities and towns may create a smaller steering committee and have consultants or system integrators manage all of the technology logistics. Larger cities could opt to have a bigger project team at the beginning stages to get things up and running, then fold the team into the nonprofit corporation. There are a lot of logistical operations to be handled by just one or two people.

Outsourcing most of the logistical work avoids having to hire permanent staff and makes things easier if the project is started by city staff and transferred to a nonprofit corporation, though there's value in having a project team that consists of several current city staff. If your first muni WiFi applications focus on improving government operations, there's a lot of politics involved with getting different departments to work together, and "insiders" sometimes can mitigate this more effectively.

Characteristics of good team players

Depending on the size of your city, the mayor may or may not play an active roll in managing the daily operations of the team. But 99.9 times out of a hundred, having the mayor be a proactive, publicly visible supporter of the project will increase your chances for success and the level of success you will achieve. To be an effective cheerleader, the mayor should have a general understanding of technology, if not be an actual user of some form of wireless besides a cell phone.

The project leader

The person tasked with leading the project team and/or steering committee should be more than just a technologist or just a politico unless most of the political forces (politicians, community leaders, business interests) are totally supportive of broadband wireless. The political landscape is too charged right now for someone who isn't deft at handling clashing interests and personalities, and the technology is changing too quickly for someone who doesn't have a solid grasps of the key aspects of wireless. However, if you can't find a techie with political savvy, lean towards political savvy and good business sense. A competent consultant can compensate for a lack of tech expertise.

Additional qualities you want to look for, particularly in the leadership role, is vision, the ability get others to share the vision, tenacity, multi-tasking capability, the ability to motivate volunteers and juggling skills. A sense of humor and appreciation for the absurd helps too. Robert McNiel, Principal with the Ronin Group which facilitated the focus groups and town meetings for the project, states that "Dianah's disarming and focused. I was blown away by her presence. There are not as many people who could pull off what she did. She's very knowledgeable about how organizations work and is good at collecting input, building momentum and then working from a positive group dynamic. She's an expert builder of collaboration rather than a 'do it alone' person. It's much easier to get things done this way."

Richard Miller, VP of Marketing and Communications at Innovation Philadelphia, a consulting group that recruits technology companies to move to Philly, says you need to look for a "rainmaker" to lead your team. "This is the person who may not be the most detail oriented, but they know how to look at something, determine what the next phase is that needs to be done and make it happen. In a lot of ways, Dianah's vision became a reality because she convinced the Mayor that this is the place to go. Finding someone who knows how to manage this kind of a complex project is also important. Dianah is a great leader when it comes to sitting down, mapping things out and then having people go do them. This is critical."

The person leading your project doesn't necessarily have to come from within the government structure. Smaller government organizations may not have enough people with this skill set that you can afford to divert from other responsibilities. Miller suggests your team leader "could be an entrepreneur. If you need someone who's knowledgeable, someone who understand the landscape, and someone who makes things happen, I would look at an entrepreneur. When Innovation Philadelphia makes investments in projects, our investments are made with serial entrepreneurs who are strong in a specific discipline, and maybe this is their second or third company. You need people with that sort of entrepreneurial passion."

The project manager

The project manager needs to be strong in business or project management skills and have some expertise in technology deployments, even if it isn't wireless technology, because success or failure really depends on how well you manage the many details. Ms. Robinson has the tech expertise plus an MBA. There is a growing trend in some industries, hospitals for one, toward having the person who is driving their wireless implementation not be an IT person, but someone with business management expertise in that industry. CEOs believe this ensures the technology applications that get built are more in line with the organization's business needs. Again, a good consultant can provide the necessary technology expertise.

Your project manager is going to handle all of the daily pieces of the implementation, including selecting the vendors and implementing physical technology components. Once these things are aligned and start to fall into place, people start asking "how can I join? Is there a benefit to my organization if we become part of this project? The project manager has to deal with these too. Varinia will tell you that often her job is a continuous juggling act.

Greg Richardson of Civitium agrees that it's a unique skill set you need for this kind of project. Any good IT person or project manager is adequate for managing the deployment phase because the requirements are pretty straightforward. But for pre-deployment there is a unique mix of business planning, economic strategy development, and policy development. There's also stakeholder analysis, planning responses to potential opposition, determining what are the level playing field issues and working with a broad range of constituent groups.

"If you need to look outside of the city government structure, obviously you want someone with a good knowledge of wireless technology, and if you can, try to find people who are also certified by the Project Management Institute," states Richardson. "This person should be a good communicator because there are so many vendor parties working on one project. Most of the bids for municipal wireless are from consortiums of several companies coming together to make one bid, so the challenge is intense."

Your project manager is going to need to be a good communicator on several levels. "For example, if you're bringing access into a community, you need to be sensitive to community and to the issues," observes Miller. "Varinia is one of those individuals who does this aspect of the job very well. She's well liked by those in the communities and there's a level of trust from them that you need to have because many people distrust technology given that it's new and outside of individuals' comfort zone."

Collectively, the team leader and the project manager have to be committed to delegating decision making and other responsibilities to all the various people and groups with which they work. Karen Archer Perry, who started the project as a community liaison from Lucent Technologies and was involved with one of the pilots, appreciated the lack of red tape. "Dianah & Varinia were very accessible, they gave us a lot of latitude to do things if we were clear about what we wanted to do and they didn't make approvals a pain to deal with. This enabled us to work very quickly."

Industry consultants are valuable team members

Whether your municipality is small or large, you may also find that a consultant such as Civitium becomes a third member of the project team leadership. This new breed of consulting firm, not to be confused with that of a pure technology consultant, blends government management and technology expertise that helps you understand and address both the business and the technical needs of your municipality. Marketing Resources, Inc (MRI), the group which helped St. Cloud, FL launch their project, is another firm in this category. These firms know the right questions you need to ask, players in the wireless industry and technology issues.

Varinia feels that, "of all the roles, the consulting group is one of our key players. Civitium has been there since the beginning and they continue to be engaged as we move to full deployment. To find the right one, you really have to look very closely at their background working with city governments and what they bring to the table." Some of the small towns such as New Haven, CT and Upper Dublin Township, PA have retained local consultants, depending on tendency to outsource. Others prefer to use internal people who learn as they go and do the best they can.

Dianah also believes in the value of a good consultant. "If you don't have the expertise on staff, which most governments don't since this is such a new technology, you need to pull in industry experts to work with you. This does add some costs to projects. In really small towns there could be a number of home-grown sources of expertise who like and play with this technology, such as universities. We tapped people from several universities to help in a consulting role to analyze various technology proposals."

How do you know if the person sitting in front of you for an interview is really qualified? Look at industry research firms such as Gartner Group, Public Technology Institute, the Yankee Group and the Wireless Internet Institute. There are many industry groups that you can look to for input. The Internet itself is great place to go and research. Also ask around to other cities that have moved at least through the RFP development stage. Most of the cities in this category such as Houston, Portland, OR and Corpus Christi, TX used consultants and likely had enough experience with them to give you a good critique of their positive and negative points.

Finally, try to find consultants who have experience with cities or counties similar to yours. Deployments can be radically different from one place to another, so you need someone who understands the dynamics and type of technology, political and business challenges you face.

The executive steering committee

Recruit the people for the steering committee from the main constituencies you expect your wireless deployment to serve. Their input will be invaluable at aligning the technology with local business and community needs. It's important that you select people who are used to rolling their sleeves up and getting work done. Philly's Executive Committee was able to do extensive qualitative research and write a business plan in just over 90 days while pilot projects were also being planned and launched. You don't get this kind of swift action from people looking to pad their civic resumes.

Dianah brought the whole group together at the beginning of September in 2004, set the objective for writing the plan, defined everyone's roles, and specified the deliverables. Everyone on the committee worked on the plan, with small groups assigned to particular sections and everyone contributing to revisions of the final document. Before people left a meeting, it was clear that they had to deliver what they promised.

When the Mayor approached the people to participate on the committee, many were not aware of the idea except a few who might have heard a brief mention from Dianah. Their reactions to the concept of municipal broadband wireless varied.

President of Talon Solutions and consultant for companies managing large construction projects, Robert Bright recalls that "first I had a sense of curiosity. What really is it? At the time, I was familiar with the technology. I knew you could have basic wireless at the house but I wasn't familiar with all of its potential uses. What does it mean, municipal wireless? It isn't until you get into it and understand some of the uses and benefits of broadband that you realize this is just the next step in wireless technology. There's not anything really new to it. You have electricity, gas and this is the next utility. Or you can look at this situation as not being very different from the Wild West, to a degree, when the railroads came out. Wireless is a new, more convenient, less expensive way to keep people connected and bring new services to people.

For Pat Renzulli, CIO of the School District of Philadelphia, the birth of Wireless Philadelphia coincided with the effort the school district already had begun, called Alternative Outreach. The District had its own robust wireless network development underway through e-Rate, a federal government program that provides funding for districts and libraries to build data and communication network infrastructure.

"The District did not see ourselves as direct beneficiary of the Philly project, though we did explore the possibility of the city taking advantage of what the schools had done," says Renzulli. "However, restraints on the funding prevented that option. On the other hand, as we experienced the benefits of our network being able to deliver more effective interactive educational content to classrooms, teachers, and principals, we saw parents as being next logical group to benefit from these materials. Wireless Philadelphia can be that last technology mile to the parents."

Ed Schwartz, president of the Institute for the Study of Civic Values which advocates citizen activism and connecting people to governments in creative ways, had a little insider knowledge of what was going on. His company is under contract to provide a lot of the Web site and e-mail services for the City's government site.

He recalls that "Dianah told me a year ago that this was an idea she had and that it would require a lot of work even though it was a volunteer job. I thought it was a great idea. I've been involved in economic development for 25 years and this seemed to be creative way to help the whole city enter the emerging digital economy. Usually Philadelphia's playing catch up, but this time we would get to take the leadership role. I certainly still believe in the plan. I think it makes sense. It's ironic that everywhere else in the country and the world except for the corporate headquarters of a handful of companies, there is tremendous interest in what we're doing. Cities elsewhere have picked up the ball. Others were doing it before Philly started, but we've given tremendous legitimacy to the idea."

One major benefit to creating such a diverse group and including people with little initial of understanding of wireless is that you significantly increase your odds of getting a plan that meets the greatest needs of the most citizens. Your main constituencies are brought into the process early so their needs and the unique opportunities that wireless offers them are factored into the plan in a more thoughtful manner. Also, these participants can start building foundations of support among these constituents you will need later. The people involved in the process who are learning about wireless are likely to spot potential flaws in the thinking of those who may be enamored with wireless, as well as broaden the team's thinking since their creativity isn't impeded by preconceived ideas.

Project teams for city department and community applications

So far we've discussed issues primarily affecting project teams focused on broadband deployment for constituent-driven uses of the technology. If you're emphasis is on deploying mobile workforce applications for use by various city departments, then there are additional points to consider.

The team leadership and project manager considerations are similar since you still have to drive the process of getting the broadband infrastructure built out. But the end users in this case are the city employees, so there are mobile devices being deployed and employees are accessing the city's business applications, intranets and possibly Web content. This is similar to deployments for commercial entities, so your project team needs to closely resemble teams in those environments.

For the leadership positions, the main question is will the business (public administrator) side or the technology side of government drive the project. Often it is IT, but many commercial organizations find that the business unit should be the main force leading the project because they know best what business operations and employee issues need to be addressed. Once they clearly defines the business needs, objectives and relevant issues, IT then needs to manage day-to-day deployment issues with the business side's oversight.

Organizations in which IT drives most of the process are those whose executives feel intimidated by anything technical or by IT's traditional "right" to run all technology projects. A danger with this approach is that a wireless deployment has greater ability than other technologies to change how an organization operates, plus it has more diverse logistical elements than typical applications. There are typically more vendors coming together, a greater variety of technologies, dissimilar end-user mobile devices needed by different types of workers, and greater application integration challenges. There are so many points of potential confusion or failure that without clear business leadership, what you get may not be close to what you want, your organization needs or employees will use.

I interviewed over 50 CIOs and business executives who've lead wireless deployments in a variety of industries for organizations of all sizes. The most common project team leaders for large scale deployments were CIOs with a high degree of business side experience and skill at getting business managers actively engaged in the deployment, or C-level business executives such as a chief operating office. Department-specific projects were pushed through by department managers. In some cases, as with Philadelphia's L & I department, their Commissioner and senior IT person have a shared business-side/tech-side leadership role. This appears to be an ideal situation when the individuals are comfortable working together.

The person or people on the team who will directly manage the implementation should be, obviously, as competent as possible. Walking down the hall and tapping the department geek is good for making sure someone from the business side with some tech savvy has influence on the project. But for wireless applications that are going to change how departments operate, you want a person at the helm with sound project management, a thorough understanding how the department and people operate and effective communication skills. It's also advisable to have the leader(s) be politically astute to overcome internal rivalries, but you can compensate for a shortcoming here by adding a few members to the team who are influential and good at consensus building.

Members of the department team

There are three important things to consider when you're putting the full team together. 1) Is the project team representative of those who will use the technology, build the technology and pay for it? 2) Will the team members in the pilot be involved with the full deployment? And 3) will all of the participants be more concerned about business objectives than getting the latest, coolest tech toys?

If each department is going to have its own applications, then you don't have to worry too much about creating one large cross departmental team to manage daily operations, but have each department create its own team. However, someone from your city's central IT staff should be involved with each department's main project team because it's likely that there will be some overlap in terms of applications or data that some of the other departments will want to share. Even if the overlap isn't apparent now, there could easily come a time when someone will want to solve a business problem that can be done more cost effectively if there aren't departmental silos of technology, meaning applications whose features or data can't be shared with other departments.

Philadelphia has the Mayor's Office of Information Systems (MOIS), which is tasked with coordinating all of the departments' IT efforts. When Dianah Neff was appointed CIO in 2001, there was virtually no centralized IT governance in place to drive projects. Now there are 460 people within MOIS and they have clear lines of authority and responsibility. IT supports 52 city departments and agencies. IT program managers oversee the relationship between IT and groupings of similar departments, such as public safety and public works.

One of their mandates is to integrate applications and databases so employees, and eventually citizens on the Web, can share relevant data from various departments. A wireless project such as the one in License & Inspections subsequently can benefit from this centralized planning body and be assured of maximum success. If your main IT department is similarly tasked, there can be one person assigned to assist the wireless project team of every department, or a different IT person assigned to each team. Either option should ensure that all of the departmental applications complement each other and maximize the use of computing resources, budgets, etc.

You can have a few people (two or three) on the project team, or a larger number. But for sanity's sake, keep the number manageable. If you can take 17 people as Dianah did and get them to turn on a dime as far as planning and launching a deployment, great. The number should be influence by how many units there are within a department so that the team's input on application design, piloting and deployment reflects the needs of those units. The project team should also include at least one person from the office staff. Almost every mobile application is going to have a profound impact on how this group does its work.

Smaller governments and other municipalities may decide to have a centralized project team manage implementation of the entire organization's mobile applications. Representing the business side leadership in this scenario might be the city manager or even the mayor, but definitely someone senior who has a handle on how all of the departments work and respect from the various managers.

Some project leaders can decide to change the team around a little after they complete the pilot project. Consider when the team is formed who should maintain their roles in the group that launches the full deployment, though the breadth of responsibilities for some of these people could change after the pilot finishes. The more senior managers may become primarily advisors rather than hands-on directors, while those with day-to-day implementation responsibilities should stay in those roles. If the pilot has been run correctly with a lot of end user feedback, it's likely the full deployment will be driven more by the IT side of the team while the business side drives the success analysis.

You add to the team's credibility if you have someone with good financial sense on board. It always helps to inspire followers and chill critics when you're meeting financial objectives and running a financially sound project. Finally, it's nice to have people involved who like playing with new technology, but you can't have the team driven by "the cool" factor. The team leader should make sure that the group is balanced between those who push the technology envelope and those grounded in business pragmatism.

Members of the community project team

Philadelphia has five formal pilot projects, several informal pilots and one or two efforts that started and fizzled because vendors who initially agreed to underwrite the cost of equipment changed their minds mid-stream. The primary pilots are: the major thoroughfare between Love Park at City Hall up to the Philadelphia Art Museum that Rocky made famous at the end of Ben Franklin Parkway; Norris Square in North Philly; the Olney section, also in North Philly; The Historical Mile which encompasses Independence Hall and the Liberty Bell; and the West Powelton section of West Philly. How many your city has will be determined by local factors and the number is likely to vary everywhere.

These initial pilots were initiated by Dianah's group who also guided the selection of members for the pilot project teams. Many cities should plan on doing the same in 2006 as there is a need for strong central direction. However, as precedents increase for the types of activities the pilots should include, communities will begin taking the lead forming project teams who work with the city to maximize use of the network.

Observations from Dianah Neff

I interviewed Dianah Neff for my previous book, "Pilots to Profits: Getting in sync with the mobile mandate." Besides providing details on how Philadelphia went about forming these pilot teams, she offers several good insights as to how you keep groups of people with various interests and competencies motivated.

Who was selected to head up the team to develop your pilot project, how, and what were their responsibilities?

We actually have five independent pilot projects going on that are coordinated through my office. The Wireless Philadelphia Program Manager was hired to coordinate the pilot managers. We have one lead vendor company per pilot. For example, HP is the lead vendor in our Historic Mile. In West Philly it's Cisco.

We chose these vendors as well as system integrators based on my research of the industry to find those who we felt were best qualified to come in and take a role. There weren't a whole lot of options a year ago, so it was a bit of a challenge. Civitium is a consulting firm with industry contacts and they helped identify additional players.

Members on the steering committee include the CIO from the Philadelphia School District, the African American, Asian and Hispanic Chambers of Commerce, the tourism industry, area universities, and the health care industry. Everything they do is tailored to the respective neighborhoods, such as helping one community project team that wants to bring in a bi-lingual portal.

Who were the other people on the team and how were they selected?

The Committee developed specific goals and established areas of responsibilities for the pilots. They relied on non-profits to help identify participants from among community residents and small businesses.

The stakeholders worked closely with the Director of the Neighborhood Transformation Initiative for the Mayor. Together they identified and classified neighborhoods by those that are in trouble or affluent. They looked for the 43% of the population who don't have Internet access. The team also got recommendations from the Planning Department which has neighborhood contacts and knows local organizations. Varinia Robinson from my office met with those who were identified and selected a team from there.

The various pilot groups did a great job defining and implementing community action programs that were facilitated by WiFi access. Some of these involved getting computers and technology education to low income and disadvantage residents. For example, the school district selected 30 families and one school in one community for whom they developed an education curriculum and provided e-mail and test programs.

Nonprofit groups targeted individuals trying to transfer from welfare to work, or moving from shelters to their own homes and used the technology to help get them on their feet. Another one is taking high-risk teens and showing them how to repair computers so the group can place computers in homes and have teens maintain them. A cardiac surgeon at Jefferson Hospital with an interest in telemedicine is exploring ways to monitor patients' condition wirelessly, give ambulances high speed connection to hospitals and enable the Center for Disease Control to send out alerts to underprivileged communities.

How effective was the mix of people on the team, and would you make any changes in the composition of the next pilot project team?

What we have are three basic coordinating groups: the School District, non-profits and the business districts within the various neighborhoods. In the beginning we actually felt that we could do a cookie cutter process and take the same pilot project team makeup into all of the neighborhoods. But each neighborhood is unique in how you need to approach people, how to use the infrastructure to get information out to people. [Philadelphia has 83 distinct neighborhoods within its 135 square-mile boundaries.]

So if you have one group of potential team members who are very concerned about health information, they can work on a team in a community where this is a big concern. Another group of people or a neighborhood might be interested primarily in small business development. Another might be interested in the bi-lingual portal. Having each neighborhood's needs shape the makeup of the pilot teams for that neighborhood is a well received process and the communities really get behind it. By default, future teams also will be unique.

There will be a major change, however, with the steering committee. We're going to set up a non profit organization to fund the overall project and make the major business decisions. It will have a separate board of nine people with wireless, financial and other business disciplines that are needed. They'll hire a CEO and the necessary staff. The City will still have influence, and community-level operations will continue the same way. The organization will use a wholesale model of contracting with vendors for products and services, then work through ISPs to deliver wireless access service.

What steps did your team take to manage executives' and managers' expectations of what the pilot would accomplish?

Our efforts in this area are consistent with general project management. We distribute reports to keep people up to date. Milestones and time frames are also published. If there are any commitments to financial resources that have to be made, we take all the appropriate steps to inform the necessary people so they know what to expect.

Anytime you have politics involved there are a greater number of people who you must communicate with and get to buy into the idea. We have a strong mayor governing structure, so the Mayor has operational authority to make decisions. This limits, to a point, how many people you need to influence at the start. I took a briefing paper on this project to Mayor Street to see if this was in line with his goals. He saw the benefits immediately and put his support behind the project. We germinated the idea within the government, then reached out and got a lot of civic groups involved.

This is not the normal process, either for many city governments or typical business operations. In many other cities, similar WiFi projects are coming up through the communities and are then taken to the government. But here, the driving force came from the government. Whichever way your organization operates, I highly recommend that you go out and talk to stakeholders early and make sure you have the support out there that you need.

Do you recommend using any particular incentives to get members of the team to work with a sense of urgency?

I think this has to come from the person who drives the team. People have been amazed at how far we've come in a year, from first pilot to where we are now with five. This is the result of keeping focused on what the end goals are. It helps that the Mayor's office has term limit because it motivates Mayor Street to push to complete this project before his term is finished.

With a project of this size, you have so many different activities that you have to have people who can drive to deadlines. As we've been successful, other cities are doing similar projects, so this is an additional motivator for us to stay first. But you don't want to move so fast that you don't do it right.

All of the participants in the pilot projects are volunteers, so there's no economic incentive you can use. They have to see that there's a benefit to the groups that they served. It took focus groups and open mic forums listening to people to generate the initial excitement. Once they come on board then you have to start an evangelism effort. As the press picks up on it, you get more motivation because people want to be part of this. Then the challenge becomes 'how do you make it real at my level?' You can't keep the discussion real high level, you have to bring the project down and make it earthy, something that people understand and support to their neighbors.

Finally, you have to let people know about the successes of the project. For example, the People's Emergency Center is using the pilot project in their area to help homeless people who are moving into transition from life on the streets into a real living situation. It's inspiring when you hear their stories, and it keeps people motivated. Once in a while you get one of these technologies that makes a significant change in people's lives. There's always risk, but there are also a lot of rewards.

The community perspective

What seems to be the norm in these constituent-driven projects is the collaborative team approach. There are constituent groups such as the Greater Philadelphia Tourism Marketing Corp (GPTMC) or community services groups such as Lutheran Children and Family Services, and these organizations partnered with leaders or other groups within their respective constituencies. Various vendors that were interested in bidding for the main deployment project, such as HP, Cisco, BelAir Networks and Lucent Technology, joined each project's partnership since the pilot was a good way to strut their technology stuff.

In a couple of cases, there were neighborhood-specific community services groups such as The People's Emergency Center in West Philly, or a larger organization such as the Philadelphia School District which already had wireless initiatives underway. In these situations, the primary task for Varinia was to bring these existing project teams into the city's efforts. The Historical Mile pilot needed to be driven primarily by Civitium since at that time there wasn't a constituent group per se to step up for this area.

There are probably as many ways to structure and populate a constituent-driven project team as there are constituencies and neighborhoods. But here are approaches that some of the groups in Philadelphia used.

Brian Loebig is the VP of Training and Quality Services for the citywide Lutheran Family and Children's Center. He played a key roll in the pilot in the Olney section. Loebig believes that one of the things most critical to the success of a project such as this is the collaboration between the community and public and private sector organizations. Tech Access PA, which provides refurbished computers to people in need, Children's Network Service, Karen Archer Perry from Lucent, Ninth Wave Media, which developed the community portal and BelAir Networks all participated in this.

"We got a grant from the Lucent Foundation for $4,000 to help roll out the program. Since sections inside St. Paul's Lutheran Church [command central for the pilot launch] can't get good WiFi access because of stone walls, we also got a bridge to put in the bell tower that boosted the signal. Karen convened meetings with people in the community and also brought in people from one of the services agency to help with getting the word out. Our Executive Director and the Director of Church and Community Partnerships were active members of the group as well."

The previous chapter described how the People's Emergency Center (PEC) partnered with the United Way and other organizations to develop a community-driven project and then worked with Wireless Philadelphia when its pilots were starting. GPTMC partnered with a vendor named Pervasive Services which created a wireless portal to help tourists in the Parkway find information and simultaneously drive business into nearby businesses and historic attractions.

General guidelines

With any project team, regardless of its focus, it cannot be stressed enough how important it is to manage expectations at all levels – politicians, employees, constituents, the media. If you look at why some teams are very successful with their projects and others have a very rocky journey or outright failure, how well they manage expectations is definitely a determining factor. As Civitium's Greg Richardson is fond of saying, "frustration is a function of expectation."

A classic example is when you have senior managers and elected representatives whose technology expertise comes primarily from James Bond movies, reading ads and listening to their new-found business flight companions. These people often expect more than a particular technology is capable of delivering. Reign in those expectations or these folks will be disappointed with the pilot results and want to kill deployment plans. At the very least, they will make it very difficult for you to get budget approval and move forward. Better to under-promise and over-deliver.

Managing the media relations effort effectively from day-one is also critical to managing expectations. It appears there were a lot of "sound bite" politics going on in the earlydaysin Philly as individuals for and against municipal broadband were making emotional statements to score points, which is what the media picks up on. And of course, sound bite begets sound bite. But it doesn't necessarily help you stay on message.

In media relations and all forms of internal and external communication, use good tactics to make sure everyone participating in and championing the project stays on the same page. Be very specific early on about what the tangible and intangible objectives are, technology capabilities, which mobile devices will be used, project timelines, costs, and the potential benefits. Put these in writing, and have all the main stakeholders signoff (literally) on the document.

Keep stakeholders at the community level informed if they are not active participants in the pilot project. After the pilot launches, any unexpected changes to the project that will significantly alter what people are expecting to see, do or achieve should be documented and distributed to the appropriate audiences. Making changes or adding new features mid-project will increase time and costs, and may even dilute the impact of an application. Be sure every stakeholder and partner involved clearly understands (also in writing) what their new expectations should be if they accept these changes.

While you want to manage as well what people on the project teams expect, don't limit their flow of creative ideas or stifle their flexibility in dealing with the unpredictable changes, challenges or opportunities that pop up. This can cripple your ability to implement the best possible application. Make sure everyone is aware of the consequences of people's decisions so there are few surprises later.

To sum it up – a project team checklist

Tim Scannell is president of Shoreline Research and columnist for Mobile Enterprise Magazine. The following is a quick list of questions you should ask yourself (or others) and some guidelines when assembling your mobile team to keep everyone on the same page. There is also some advice here for working with contract IT staff and consultants.

1. In planning for the initial pilot project and launch of a mobile project, have you created a budget that takes into consideration the costs associated with using people inside the organization, in addition to contracting for services outside of your organization? Often, it is a smarter and more cost-effective move to pull in talent from the outside when needed, instead of dedicating too many internal people who may spend a lot of time sitting on their hands.

2. Are systems established that allow team members both inside and outside the company to share ideas and track progress during a mobile project? If not, then

there might be a lot of duplicated effort. Worse, their efforts may not be in sync as the project progresses. A good idea is to decide on a single tracking or milestone application that can be used to see the general picture of activities. Network administrators might also set up a dedicated chat or 'idea sharing' area on the email server, or even a separate Web portal that is focused on mobile initiatives.

3. Do you have the right mix of people involved in your mobile project? Obviously IT and managers from the business side will be involved. However, it is also important to include in the initial planning sessions potential users of the mobile system. It is also a good idea to familiarize these people with client devices that may be used on the network and even let them have a vote in selecting these devices.

4. Information-sharing is important when testing and launching a mobile project. However, it can be a nightmare if you have not made the necessary technology and security provisions to handle outside contractors. This means establishing secure accounts and authorization levels for contracted workers, and making sure they fully understand your internal policies and procedures as they relate to what can and what cannot be done over your wired and wireless networks.

Be sure your contractors are well aware of exactly how much security you want to layer on top of your wireless system, and how you want that security balanced with ease-of-use and other considerations. Obviously, a college university is more concerned with general access and widespread reliability than triple-DES security and encrypted files. By the same token, a business that does contract work for the city government would stress security over ease-of-use issues.

5. Hiring talented people is important, but it is just as critical to pull together the right team personalities when it comes to working on a project that may require long hours and bring a lot of different egos together under the same umbrella. It is a good idea to avoid the 'code cowboys' who think mobile programming requires shoot-from-the-hip approaches and a minimum of team coordination. This will not be successful now that mobile solutions are an inseparable part of the enterprise network family.

6. Accessibility to team members, wherever and whenever is needed, is critical during the early 'make or break' phases of a mobile project. Therefore, it is crucial that all internal and external staff members adopt an 'ASAP attitude' when interacting with each other and responding to questions and demands. A typical mobile pilot project might involve dozens of end user participants, many of whom have no previ-

ous experience with mobile systems or networks. So, some of the questions might be embarrassingly simple, or have been asked 20 times before. Make sure all queries are answered, and problems resolved. It is also a good idea to issue interactive mobile devices such as smartphones to team members to keep everyone in the loop.

7. When interviewing outside staff for inside mobile project jobs, be sure the prospective candidates know as much about your basic business as they do about mobile technology and applications. Far too often, people are hired who know all about the technology, but may fail to successfully apply and use it in highly-targeted situations. For example, wireless access on a city street may essentially be the same thing as wireless access in an office. However, there are different security and user demands that affect the ultimate design and operation of the network.

8. Nothing breeds disaster like ignorance, so be sure to keep everyone who might eventually be involved as users in the wireless network informed of what you are doing and what your general goals are in deploying a wireless network. The early stages of a project can be a little hectic, as people on and off staff walk throughout a department or building with wireless routers and RF signal sniffers to map out the best configuration. Telling the general population what you are doing will prevent someone from rearranging the metal furniture after you've done a site survey. An Open Door policy may also lead to some worthwhile suggestions from the mobile user community.

9. Will your wireless access for outsourced talent end when the final application is up and running? It would be a serious mistake not to cancel user access and passwords for contractors who no longer work on the project or the company. But, it might be a good idea to extend remote access to contracted workers through a mobile portal of some sort to perhaps continue the idea sharing and add a touch of continuity to the project. These outsiders wouldn't have access to the central resource, of course, but they might be contained in a separate region for additional comments and input. Systems integrators might also consider launching a mobile site or portal to offer continued training on mobile applications, and keep a foot in the door for future changes and enhancements to the network.

Now, let's move on to the discussion on consensus building. As with managing expectations, if you don't do this part of the job well, there will be serious wailing and gnashing of teeth to contend with sooner or later. Probably sooner.

Chapter 5

Team Philadelphia –
a consensus-building machine

Consensus building is something you need to plan for from the first moment you broach the subject of municipal broadband. Some of you already know what I'm talking about from first-hand experience. There likely are some rough moments ahead. Your biggest political battles may be between departments, not at the state capital. The larger the municipality or county, the more constituent groups there are with different needs and agendas. The topics around which you must build consensus can shift between this month and next.

Everyone working directly on the Philadelphia project and others familiar with it will tell you that this wireless initiative owes a lot of its success to Dianah Neff's ability to get people with divergent interests working from the same page, or at least in the same chapter. Her abilities were certainly tested given that she and her team found that the amount of time they would eventually spend consensus building among a wide ranging network of community, business, political and academic groups was staggering.

A great deal of the success of your consensus building efforts rests on the skills of the person driving the effort to build bridges, form alliances and continually broaden the circle of believers. The rest depends on the project team and steering committee members each being a leader in their own right, carrying the message and working the crowds (figuratively and sometimes literally).

This chapter shows what went into the process in Philadelphia: Dianah's meeting with Mayor Street to present the idea of muni WiFi; getting the Executive Committee on board and in sync; reaching out to diverse constituent groups; responding to the political challenges at both the local and state level. Even though it all looks daunting, it's still manageable.

What you say and how you say it holds the key to consensus building

In the early days of the Philly project, the real meat of many issues often got buried in fiery blasts of rhetoric from certain quarters rather than providing the food for thoughtful discussions and working together. Luckily, the Executive Committee and the project's many supporters managed to rise above the noise to build the necessary consensus to keep it moving, though some of the contentious battles will never really be over until the city's network finally goes live.

Here are several points and recommendations to help shape messages that rally people around a technology that's a little difficult for some of them to fully comprehend. There's always more you can do to ensure that your early stages of consensus building are successful, so the following is only a starter kit for now.

First, whatever you mission is, state it consistently. Wireless Philadelphia has always been on point, describing the project's social and economic objectives in its business plan, Web site and other communication. If your municipality is going to improve government operations first with muni WiFi and then tackle social issues once the infrastructure is in place, be equally as clear. Forget lofty statements. You're not writing the next Declaration of Independence. Rely on simple and to the point, such as "We're going to use wireless broadband to create a more efficient, responsive government." Clarity breeds consensus.

Second, resolve the business model quickly and don't talk about it publicly until you do since this is where roadblocks to consensus building can pop up. If you have no intention of going into the wireless service business, state it starting today. If you plan to create a government/business partnership to offer broadband access services in which the city or county facilitate infrastructure access and service providers provide all of the business logistics of selling and supporting the service, lay it out there. Haven't decided yet which model makes sense until you talk to constituents? Then that's your public statement. The clearer you are on these details, the faster you build consensus. Keep this discussion focused so you minimize misconceptions about your model.

Third, make sure people understand that municipal wireless broadband (muni WiFi) is not DSL. Some critics attack muni WiFi plans by saying "we already have DSL and it's just as good as WiFi." DSL is high speed Internet access service that comes into a home or business through a fixed wire connection. Muni WiFi is high speed Internet access without the wires. People using it can have just one account and get on to the Net from their homes, in the street, in their offices or anywhere else the network covers. This un-wired access is what enables many of the benefits described in Chapter 2. Use the analogy of "portable DSL" if it helps people understand.

Fourth, make sure you have the facts about your community that help create a sense of urgency to make this project work. The U.S. has dropped from 3rd to 16th in the world in terms of broadband adoption by citizens, and the quality of our broadband is way behind many other countries. State how this impacts your business community that has to compete in a global economy, or disadvantaged businesses that could grow by opening new global markets to stay ahead of local competitors. A recent U.S. government study reveals that 11 million people across all socioeconomic strata in America are illiterate. Relate this to the numbers in your community, and how your WiFi initiative can help reverse the trend.

Fifth, be sure people understand that the economics and deployment logistics of broadband wireless offer advantages over alternative options for residents, businesses and other organizations. Tell them to check the fine print to do the real cost benefit analysis. What will "special low prices" revert to after the promotional period ends? When companies advertise data speeds of between 400 kbps and 700 bps a year, is that true speed or just the speed to download files. When built out properly, WiFi exceeds these limitations going and coming.

Include these along with other performance, price and logistical advantages of WiFi, relating everything back to the mission statement. For example, if your mission is to tackle the social issues of the digital divide, presenting consumer-targeted price and performance facts is important for winning over opponents. Or if your main objective is economic development for disadvantaged businesses, show that WiFi isn't just about low cost technology; it also increases productivity for these companies. As John Dolmetsch, president of Business Information Group in York, PA, says, "when you sign up for a DSL connection, it's working only at your house. When you get broadband wireless access, you get the ability to go anywhere and work from anywhere."

Sixth, find a tech person who can talk about technology so that the person with low technology-comprehension skills understands. Have him or her write one-page overviews of the various components of broadband networks and the applications you plan to deploy. Several people I spoke with who are pretty smart and wholeheartedly support the initiative were nevertheless confused by the basics of wireless in their first meetings with representatives from the city. Without clarity when describing what the technology does, recruiting support is challenging as you reach out to the constituent leaders who need to make wireless projects work in their neighborhoods and conference rooms.

Creative, occasionally comical, but never condescending materials are the way to achieve that clarity. Johan Kroes, Senior IT Architect for a financial services company in South Africa called Sanlam, states that "a great part of my job is to explain in simple laymen's terms what a system can and can't do. Make little analogies – this application is like a 4 x 4 jeep. It's great for these tasks, but you can't race it in the Grand Prix. Or now that you have a Grand Prix auto, do you want to take it on the beach? This connection is like a water pipe in your sink. You can't force 1000 gallons of water through here quickly." Make sure the people speaking on the project's behalf avoid jargon and geek speak.

After shaping the messages to win people over to your corner, the hard work is executing the activities that rally the various individuals and organizations around the messages so they take action that builds the broad support you need to move forward.

Philadelphia puts its ducks in order

The main person to initiate and sustain the opening push to recruit supporters for the wireless initiative - either muni WiFi or department-level applications - may not be the same person to head the steering committee and project team. It could be anyone who has a vision plus an aptitude and interest in learning how wireless can benefit governments. This person needs to be committed to doing a lot of leg work researching general wireless technology and specific potential products for the city so he or she can make a credible case and turn everything over to the eventual project leader. They also should accumulate basic financial data about what these technology options may cost as well as what benefits the city and its stakeholders can reap from the implementation.

Prior to becoming the City of Philadelphia's CIO in 2001, Dianah had undertaken a similar municipal broadband project in San Mateo, CA. Even though Mayor Street did not ask her about the municipal WiFi until 2004, she never stopped researching the possibilities. "I had been following wireless LANs for quite a while and took note when the FCC apportioned a part of spectrum that was unregulated because without this we wouldn't have been able to move forward," she states.

Then the IEEE (a group that sets standards for various technologies) approved standards for 802.11b, 802.11a, which is important for indoor wireless access, and 802.11g which facilitates longer data throughput. These standards enable the various infrastructure and hardware products needed for the network to work together and with future products that support the same standards, thus protecting cities' investments from future changes in the technology. Dianah continues, "when we saw that the number of devices such as laptops, PDAs and wireless cards with embedded WiFi chips would hit the billion mark in 2005, that was an important milestone. If you are going to have ubiquity across the city, you have to have affordable devices and they have to be in widespread use."

This preliminary work doesn't have to be a formalized process, nor does someone need to be "officially" recognized to pursue this task, as was the case in Philly. In plenty of companies it's the lone wolf working in stealth mode who does it and when they feel they're ready, will take the idea to the powers that be. So once the legwork is completed, the muni WiFi torch bearer needs to make the first pitch for consensus to the most senior manager or elected official who can give the project legs, and political support to sustain the effort once it picks up momentum.

It may be useful to gather a little informal feedback from constituents. While she was accumulating necessary technical background information, Dianah also put out some feelers locally. She talked to community players to see what they thought about these ideas, particularly the school district and the universities. Everyone was very supportive of the concept of municipal broadband. This plus the changes in the industry influenced Mayor Street to ask Dianah in April to go out and test the idea in a proof-of-concept pilot. It was wildly successful. In two months, and without much publicity from the city, 1200 people signed on to use wireless in Philadelphia's Love Park.

Building consensus within city hall

Many technology initiatives in companies fall short of their potential benefits to the organizations because of the lack of vision and direction from top executives. If you do not have the enthusiastic buy-in for a mobile or wireless implementation from the senior executive team, particularly the president and the person who approves the checks, successes will be limited. The company won't reap the full potential that wireless technologies offer.

Many of the organizations producing impressive results with wireless applications are doing so specifically because they have that top-level support. Once mid-level managers see that this has become a priority for their bosses, they tend to fall in line to ensure widespread acceptance and cooperation from the rest of the organization. The necessary resources are committed, plus you get the extra high-powered backup when the inevitable roadblocks to progress pop up.

The same dynamic holds true in government. The operational structure of government varies by municipality and county. But regardless of whether you have a strong city manager, a strong mayor, part-time elected officials or appointed people running the day-to-day operations, you have to get public support for wireless from the main person at the top who makes things happen. In some places, the person who carries the clout may be one of the senior council members. Know where the power lies and cultivate it. This is also true at the department level. It is possible to launch an initiative without this support, but the chances for success are not great.

The Mayor steps into the breach

"Mayor Street was re-elected in 2003," states Neff who, as CIO, is also a member of the Mayor's cabinet. "We looked at what were the goals in this administration, what were the major projects within the city that we wanted to complete. Also, what were those projects that could be started and finished during his administration? I brought up wireless as one of those, and was able to tie that into neighborhood transformation, which had been a cornerstone of his administration since 2000."

It was a natural progression. The city was moving abandoned cars off of the streets, closing drug corners, greening the neighborhoods. Now they needed to bring in economic vitality and overcome the digital divide in those neighborhoods. Wireless was the technology that could meet these goals. This really wasn't a hard sell for Philly's top tech, given the Mayor was particularly predisposed to an idea such as this.

Recalling his election campaign, he states "I was headed to one of those community forums. My opponent spoke before me and he said 'if I become Mayor, I'm going to have mini-City Halls in every council district because people shouldn't have to come to Center City to pay their bills.' After I spoke, someone asked me what I thought about that, and I said I don't like it. I think it's a bad idea. The future of this city is online, not in line. The room lit up because people got it. I personally made a commitment there that we were going to be a 21st Century world-class city. That we are going to lead the way in the area of technology, both as an integral part of the way we do business as a city, and in the kind of environment that we create for the people who live here, work here and visit here."

So as he listened to Dianah's briefing, he grasped how broadband should be the lead technology to bring about the transformation he desired. He became a believer. Later at a conference, Mayor Street would reflect on a story he read about a small town with 40,000 people as opposed to Philly's 1.5 million. At one point they had five homicides in a seven-day period. He looked at their response to this. They were cleaning up streets, boarding up abandoned buildings. They were doing similar things Philadelphia was doing because both cities' challenges were fundamentally the same. This prompted a thought.

"We think things at one level are going well because we have all of these programs in place," he says. "But these things alone won't do it. Clean streets and affordable housing by themselves will not create the bright future that we have for our city. We believe that the future of our city is going to have everything to do with the use of wireless, broadband technology."

Setting Philadelphia's wireless efforts in motion was also helped by having a mayor who is committed to personally using the technology. He's is a self-acknowledged gadget freak who carries a couple of different devices with him everywhere. In quite a few commercial operations, part of IT and business managers' strategy for getting senior-level support includes giving all of the top executives PDAs. This may or may not be practical or even necessary within your organization in order to get their sign off on broadband.

However, if your senior officials or department managers lack this hands-on experience, show them in stark terms what wireless can do for employees and constituents. Draw clear lines from the inefficiencies of the current ways of doing business and providing services to ROI that is made possible by WiFi. Consider taking a video camera along for a ride with two or three of your field staff who really care about what they do and want to make things better. They'll show you where the problems are. Bring back visual evidence for the leadership that makes your point crystal clear.

Early leadership can take many shapes

In many small towns and some cities the mayor may be the major initial catalyst. Or the mayor together with two or three other people drive the process from the start as was the case in St. Cloud, FL where the Mayor, City manager and CIO put their plan in motion to provide free wireless access for every citizen and business in town.

"The thing that got me going was a program our school district has for underprivileged children who can't afford computers at home," recalls Mayor Glenn Sangiovani. "The district gave each of them a laptop so they could do their homework like the other children. I thought if they couldn't afford the computer at home, it was possible they couldn't afford the connection either. I look at this wireless project as another way of closing that digital divide. What good does giving them the laptop do if they can't get on the Net? When we analyzed the numbers we discovered that $4 million annually for Internet access charges was heading outside of the city from all of our residents. You look at social impact together with the economic boost that putting money back in people's pocket is going to give us, municipal wireless was worth investigating."

Within county governments level, who will drive things varies as much as the types of governing structures. Some counties have an administrator who operates similar to a city mayor, while others have Boards of Commissioners varying in size from three to more than 25 members. These could be elected or appointed officials.

Jeff Arnold, Deputy Director of Legislative Affairs for the National Association of Counties and a member of its Telecommunications and Technology Committee, believes that there needs to be someone assertive who commands a lot of respect taking the activist leadership role. However, who will start the ball rolling is uncertain.

"I think what ends up happening is that the community actually starts the process. A citizen or group of citizens insists this is a service they need. But more often than not they'll just say 'we need broadband.' Either that county's administrator or a member of the Board of Commissioners has to step up to the plate and say this is something we really want. It's all about someone understanding the nature of the wireless solution and what it could mean, and being able to articulate this in specific terms. It's one thing to say 'wireless broadband.' But what does that really mean? Is it WiFi, is it fiber? And what happens if people don't have it? If people haven't experienced it, how do you get them to understand what it can mean from a business standpoint as well as a personal one."

Department level consensus building presents challenges

As for departments within city or county governments, the general dynamics of getting a wireless application project started are the same with a broadband initiative. Either someone on staff believes that wireless can help the department, to which end they do preliminary research and maybe some development work to prove the idea, or the person at the top of the department or the city's hierarchy gets the vision. For Philly's L & I Department, one of Acting Commissioner Solvibil's predecessors provided the spark for their wireless effort.

"Fran Egan was the first Commissioner to put it on the agenda to get us wireless," he states. "She started us down the path. Under her guidance she hired an IT director and put together an IT staff that Jim Weiss now heads. She used to get in the Managing Director's face and insist 'this is what we need to do a good job, so this is what we want!' If we didn't have that foundation she started five years ago, we never would have gotten to where we are."

There are many examples of city departments jumping into the wireless waters. However, quite a few muni WiFi champions run into obstacles when they're driving to bring the entire government structure onto the network.

Even though there are a lot of forward-thinking department leaders, for example, are they tenured employees or political appointees of the mayor? As forcefully as a mayor might drive the move to wireless, there's a layer of management that's tenured and will likely be in those positions after mayors and their appointees have moved on. Some folks in this layer may not share your enthusiasm for new technology and hope they can stall until a new administration is sworn in. Add to that the nature of departmental turf warfare when it comes to budgets, and you have a charged environment that requires deft political skills.

Craig Newman, Director of Business Development for Motorola's Canopy Wireless Broadband business unit, spends a lot of time meeting with cities. A common problem he finds is that you have a business culture in which, if you try something new and screw up, you've got that label for life. It will take the longest time to bring in a new idea if someone's been burned. For some people, the trick to get them on board is to guarantee them the equivalent to the legal world's immunity from prosecution. Put a certain amount of onus for the success of this initiative on the city's project team. Assure managers that they're not going to stand alone on this endeavor.

Of course, this is only part of the approach. You need to find directors' resistance points and figure out how to neutralize them. Public safety directors whose responsibilities include emergency response operations are pretty conservative and the idea of their operations being run on a shared network is a foreign concept to many of them. But budget pressures brought on by economic issues, particularly homeland security demands, are increasing the pressure to quickly find new ways to improve several serious communication shortcomings.

"Public works guys are all utility people for the most part," remarks Newman. "From their perspective, they can't pursue initiatives that raise rates because there'll be a public outcry. In a sense, it's about responding to what the voters want. If there's no incentive or outcry from residents to make things more efficient, then you do just what's good enough to make sure systems work. Also, a lot of these guys don't trust vendors because they've been burned in the past."

The solution therefore is to present ways in which the WiFi infrastructure allows the various departments to enhance their operations for less pain and risk than the departments striking out on their own. Many police officers are riding around with rugged laptops that also happen to have WiFi cards. Show them the easy bridge from current proprietary or cellular networks to faster, less expensive WiFi technology.

"Every department of parking services dreams about a wireless parking meter solution, but they can't cost justify putting the network in place on their own," adds Newman. "Municipal WiFi produces the necessary missing link. Tell managers 'you can do all those things you want to do because the city's picking up the costs for the infrastructure and the access.'" Now your initiative moves from being a potential budget encroachment fight and inter-department tug of war to the early arrival of Santa Claus.

Along these lines, in St Could, FL their mayor brought all of the department directors together, gave them a detailed briefing on the possibilities of wireless and sent them back to meet with their staffs. Each department had to come up with a least one idea of how the technology could specifically help their department.

This effectively got the directors engaged by having them drive the brainstorming for their respective areas so they had some ownership in the project, and they were involved right from the beginning through every step of the process. It wasn't something that was dictated by the mayor. They and their staffs saw the value of what this technology can do for them, they helped build the business case and they had greater influence in shaping the direction of deployment, all of which got them excited about making the deployment a success.

Don't forget the rank and file employees

"The project teams that develop these applications within a business or organization are generally comprised of executives, mid-level management and IT workers," states Ralph Nichols, Service Program Manager for the Document Messaging Technology Division of Pitney Bowes, who has extensive experience deploying mobile and wireless applications. "These are supposed to be the people who see the big picture. They actually see just a part of the picture. They usually have little understanding of the mobile environment, work-life issues and hassles that remote and transactional workers deal with daily. Many have never had a field assignment."

No less important than getting top-level buy-in is the consensus building you do with rank-and-file employees when it comes to deploying mobile workforce applications. These employees hold the answer to wireless ROI because if they are resistant to the technology, they won't use it or won't use it well. To ensure success in getting workers' consensus on your technology vision, give them what they want. More importantly, give them what they need. The only way to do that is to bring them onboard with the planning early and keep them engaged in the pilot projects.

Interviews I've conducted in the past few years with line managers in government organizations confirm that their employees are just as interested as their commercial enterprise counterparts in improving how work gets done. But employees also need to see benefits from wireless that are important to them. Some benefits come from process improvements that save them time, or procedures that are made simpler and easier to execute using wireless in their work environments. New process designs for wireless that fail to seriously consider the work environment where they're to be used don't produce positive results.

Consensus building is ongoing, so as you proceed with implementing your pilot project plans, be sure the project team members running the pilots are attentive so they detect and resolve the first signs of end user resistance to the application.

Building consensus within the project team

On the face of it, getting your project team for internal applications on the same page is a straight forward proposition. You bring them in the room, spend some time explaining the vision, assure yourself that they're buying into the vision, give them their marching orders and away everyone goes. However, consensus building for the project team driving the citywide network deployment can present some challenges.

Mayor Street and Dianah Neff did double duty getting the Executive Committee on the same page. The Mayor took care of the first round when he recruited everyone. He and Dianah had worked out the mission statement, so each member was brought on the Committee because they believed in the mission.

The next round was Dianah's as she worked to get 17 people to come to consensus on a business plan that defined a network about which there were many unknowns, and created a business model for which there weren't clear precedents. If that wasn't enough, this plan had to be clearly articulated so Dianah could then build consensus around it from the various constituent groups. The Committee as a body wouldn't be there to help since they were disbanded after presenting the plan to Mayor Street.

If the communities your network serves have diverse constituent groups with varying and sometimes competing needs, interests and goals, the project team or steering committee by default needs to be similarly diverse. This is the best way to bring the perspectives and feedback to the table to ensure that what you propose to your citizens are the tech solutions that they will support. With this diversity on the committee, however, you add more layers of difficulty getting everyone to support a document as complex as a business plan.

Veronica Wentz, Web Site and New Media Director for the Greater Philadelphia Tourism Marketing Corp. and member of the Executive Committee, states that "everybody was coming from a different place, and not everyone had the same level of understanding about the technology initially. Dianah did a good job of educating the group on what wireless broadband was about and what it can do. Then she asked people to bring back ideas on how they thought the technology could be applied in their business or community."

It helped a great deal that the People's Emergency Center (PEC) had a community WiFi project up and running that was brought into the Wireless Philadelphia series of pilots because Dianah was able to use this to make the mission less abstract. Wentz continues, "she always presented things to us to show how the technology had already made a big difference in some people's lives. 'PEC helps people better integrate into society. They refurbish old computers and have people up and running on the Net. This woman who was handicapped and had kids was able to start a business in her home and she pays $5 a month for wireless service.'"

Dianah gave very good examples of how, in a small and focused way, this technology was being used. It was really easy to gain support around this. She kept everyone focused on the vision.

Another factor contributing to getting this diverse group in sync was "the fact that the mayor had put out the charge, 'we're going to do this." adds Pat Renzulli, CIO of the School Board of Philadelphia and also a Committee member. "It was more an issue of how we did it rather than if we did it. I don't recall that there were many contentious issues. From the outset there were a lot of interviews of constituencies. What these tended to do was increase the reasons why we wanted to launch this project."

Richard Miller of Innovation Philadelphia concurs. "The discussion level in the group was always collegial versus controversial which helped us be successful. The blend of the people made this work. We all donated a great deal of expertise from our various fields – marketing, financial, technical. People were visionary and we're used to executing. Our reality was, here's the project, here's the concept. Now, how do you go through the process of getting it done? Who you need on the team are people driven by the passion that comes with being able to take a concept and build that concept into a reality."

Feedback from focus groups which ran concurrently with the business plan writing gave the group even greater determination to make muni WiFi work. It also helped them write a plan with greater persuasive powers in uniting people besides the committee behind the project. Veronica observes that "after learning more about what the community felt, Dianah was able to bring these results into the discussion. She was able to say 'this is how it's being received,' 'this is how people see applying it' giving us more proof the idea was good. You need to pick out key points that by default generate universal support and then push these forward to the team and then to the communities."

Consensus building among constituents

The population size and degree of diversity within your municipality will dictate the breadth of your consensus-building efforts beyond City Hall, while local politics will determine the intensity of your efforts. In cities with a notable range of professions, economic strata, ethnic groups and ages, you should spend time meeting with a lot of groups to get their buy-in initially, to shape the direction of community-focused projects and to keep them updated.

There is a significant risk here as the steering committee – and later the project team managing the network deployment and oversight – tries to establish its vision while helping constituencies refine their respective visions. The grand objective must align with these divergent objectives or else the programs, online content and applications you build for the network will fall short of their potential. For an implementation as visible and controversial as muni WiFi, many constituents won't give you a second chance to prove yourself if you don't get it right the first time. You can build it, but people don't have to come.

Wireless Philadelphia's masterstroke was its 20 focus groups, each one representing a key constituency that collectively gave fair representation to technical, business, service and neighborhood interests. Two subsequent town meetings pulled in a greater numbers of city residents to both learn and provide feedback. These events got the word out to the communities that something exciting was getting ready to sweep Philly. People may not have understood it fully, but they wanted to be at the station when this train came in. "You can really go by what Philadelphia did because Philly did a lot of things right," states Robert McNeil, Principal of the Ronin Group which facilitated the focus groups and town meetings.

With the business plan completed at the end of 2004 and reflecting the wishes of the communities, Wireless Philadelphia moved into the next phase of consensus building. This relied heavily on pilot projects that got constituents in high-profile neighborhoods actively using muni WiFi so the rest of the city would be more likely to support the initiative. The Executive Committee was disbanded and Dianah's remaining project group took on full responsibility for technology deployment. Through Varinia Robinson's efforts, the team still built support for the project, but community leaders became the main drivers for census building within specific neighborhoods, among businesses, on college campuses and within the health care community.

Smaller towns and cities have it easier

Little towns and rural areas tend to be more homogenous than big cities, so you'll likely only have two or three main constituencies to address and subsequently a lot less work to do. The objectives that you try to get people to coalesce around can be very different than what you've read about so far. Do your consensus building early so you get this heavy lifting out of the way, or conversely, find out quickly that the citizens won't support the project. By the time you complete your business plan, your community supporters should be fully on board.

An important thing for small cities and towns to remember is that the Philadelphia's story is one of "We the People," which makes sense given the city's history. As Executive Committee member Robert Bright describes it, "Our approach reflects the characteristics of Philly in terms of demographics and the Mayor's neighborhoods initiative. 'We the People' have an ambitious agenda. 'We the People' want to make the school district better. 'We the People' want to enhance the city to make it attractive for business." When you go about consensus building at the community level, you can pursue some activities similar to those Philadelphia did, but modified to reflect your area and the particular aspects of its respective constituencies.

Philly's strategies may be similar to ones required in smaller cities, such as in the area of economic development, but these municipalities may not want to execute tactics on such a large scale. Smaller towns may have higher average education levels while rural areas might have lower levels. The rate of poverty is different. All those digital divide issues could be there, but in some rural areas it may be due to geographic isolation rather than income. And of course, smaller communities don't draw the same type of political fire. If Kalamazoo, MI decides to own its own wireless network, it's likely none of the incumbents will go out of their way to object.

In the Philly trenches for consensus

In April of 2005, the city announced the business plan, formulation of the non-profit corporation (Wireless Philadelphia) and release of the RFP for the citywide network. Infrastructure was deployed for pilot projects in several tourist areas and communities. These would not only test the capabilities of various vendors' products, but also give the city high visibility platforms that influenced consensus building throughout the city.

One of the vendors bidding on the initiative's RFP was Lucent Technology, so they offered to donate equipment for one of the pilots. In May their community liaison Karen Archer Perry went knocking on doors in the Olney community of North Philly on behalf of Wireless Philadelphia. Her objective was to start building collaborative projects with organizations such as the Lutheran Family and Children's Center.

Brian Loebig, VP of Training and Quality Services for the group, states that "our mission is to help people in need, which dovetails with the city's mission to get technology into the hands of people who need it. To be successful throughout the city, the city needs to continue to form these types of partnerships. If community people understand the services that are available to them, it will help take them out of poverty. Our organization sees the Net as an important way to do this, and this wireless initiative as an extension of the Net's capability. But you also need to have residents in the neighborhood take ownership of the project so they don't view it as someone coming in to do it for them."

It is particularly important that you approach this process without a lot of preconceived notions. "In some other neighborhood where a lot of the women stay at home, online Martha Stewart recipes might be the thing to have," Patricia DeCarlo, Executive Director of The Norris Square Civic Association observes. "But that would be in that neighborhood. There were initial thoughts from the City that people could sit in the park here in Norris Square with laptops, which is very middle class yuppie. Our folks do not have little laptops that they can take to the park. As this wireless becomes more real, people become more vocal about what they need to do to make it better. Listen to them. Otherwise you're just wasting people's time."

Business communities may require different tactics

Veronica's organization is pursuing similar efforts with businesses and visitor attraction organizations whose revenues depend on more effective communication with tourists and other visitors. She sees the muni WiFi concept being embraced by quite a few establishments in the restaurant and hospitality industry. As for the business community at large, there are differing opinions on how much you do to get them supporting of the initiative.

"Some say that it's the same type of consensus building in the general community as for businesses in an underprivileged area, but I think it's a little bit different," says Robert Bright, who was appointed to Wireless Philadelphia's Board of Directors. "Would it be nice to tell businesses 'these are the benefits of WiFi, this is what it means, over there is such and such?' Sure, it would be helpful if someone shows you a few things. It's also possible, as a business owner, to learn some things on your own doing research."

For merchants whose neighborhoods are their business world, extra efforts to bring these owners on board with wireless would be helpful. But business owners need to understand their customers. If they're moving to wireless then companies need to go that way. Robert adds, "maybe I'm biased because my clients are not mom & pop operations, so I have a different need. When my clients say 'we're going to do all payments electronically now,' then I'm going to have to figure out how to do business like that. So there has to be some level of owner responsibility depending on whether or not you're serious about your business."

An opposite viewpoint comes from those who are directly committed to the economic development and improvement of the city. They believe that that this muni WiFi initiative will create additional jobs, enable businesses to expand nationally and globally through the use of the technology, and make Philadelphia a more competitive market in which to do business. Richard Miller advocates "an education process through a marketing campaign within the city, and also to the world, that shows small business owners the benefits, and how they can expand in ways they never thought of before. This education makes the use of the technology more effective."

How aggressively you build support from commercial interests depends on how vested the project team is in your economic development effort, whether the discussion is just about businesses in blighted areas or the entire business community. I contend that, if you believe in municipal WiFi, go full out wooing the entire community. Here is where a lot of money, and therefore political clout, rests. It's easier to keep political opponents at bay at the city and state legislative levels if you have strong business support. Of course, if like Philly, one of the main businesses headquartered in your city is an incumbent such as Comcast, working this tactic becomes a harder row to hoe. You have to play the hand you're dealt.

General guidelines for constituent consensus building

From City Hall to the pool hall, until you know what people want, don't try telling them what they need. That's not consensus building. Let them tell you what they need or want. Focus your public proclamations on key milestones as you reach them, and relate these milestones back to addressing the input you get from constituents.

Think globally about your city, but act locally. Make sure you have a good system in place to quickly identify, recruit and mobilize neighborhood and business champions for the project. Assign people to where they are needed most. The bigger the municipality or county and the greater the challenge, the more champions you need out there bringing the various constituencies into the cause. When all is said and done, when the network is finally deployed, getting citywide usage is going to be a neighborhood by neighborhood effort.

The biggest threat to the success of community relations campaigns supporting this initiative is inaction. Some people will talk or study an issue to death. At some point all of the various decision makers have to shut up and push the "Start" button on building public support. There's never going to be the perfect technology, political climate, business climate or product price.

Diary of a Community Liaison

From her experiences Karen has distilled several guidelines for maximizing your efforts at building consensus at the constituent level. She is now working under a new boss – herself – as Principal Consultant for Karacomm, specializing in communications and community relations consulting.

Meet people one-on-one. The first step in community consensus building is actually to engage people one-on-one and in small groups to tell them about the upcoming service, answer their questions and ask them about their needs and the needs of their neighbors. People's ability to get over the technology hurdle and adopt something new requires a personalized view of how it will help them.

Clearly written collateral material is a must. Pay close attention to creating well-designed frequently asked questions (FAQ's), brochures, maps of potential service areas, and other information that will allow people to understand how they will access the network. If people read good information that's localized for their community, their interest level and support for the initiative increases significantly.

Municipal wireless is not a one-size-fits-all offer. What people want to do with the technology dictates everything from end user and on-premise equipment to what they're willing to pay for it. You and the vendors and service providers you partner with must know and be able to explain what technology options best fit which needs.

Identify early adopters. Nothing sells better than local a reference. Use initial meetings to identify early adopters who see the value in the program and who want to champion the program more broadly in their community. They are the ambassadors for municipal wireless. Look for diversity in initial contacts: such as people from area businesses, church groups, and different cultural groups that will share their enthusiasm with neighbors and associates.

Make the wireless network tangible with launch events or demonstrations. If you are not already surfing at high speed, it's difficult to imagine what it means to have the world of information at your fingertips over the airwaves. Design a demonstration or launch event that showcases both the technology and the content. These must go beyond answering questions about technology, service and pricing. Constituents must get hands-on learning experiences to see what is available to those seeking information, services, education and entertainment, and the speed at which it is accessible.

Plan mini events. While a large launch event or demonstration is a great way to introduce the new service, small activities are good for on-going efforts to reach people and build support. Consider attending chamber of commerce meetings, back-to-school nights, community meetings or even doing something on a street corner that shows off WiFi service. This is possible even if the service is running from a local store or café while you build out a proof-of-concept or pilot network.

Capitalize on interest with a local portal. If you have the resources, create a basic wirelessly accessible community Web portal that links to neighborhood and business groups and includes local news or events. This enables people to get a more personalized feel for what muni WiFi means to them.

Stay connected. Circle back occasionally and check in with these constituents, particularly the early adopters. Continue to share information with them as well as learn from them how to improve your plans and reach more people.

Building political consensus

The most dramatic, near hear-stopping moment in the project came in mid-November, 2004. The Pennsylvania state legislature breathed life into a telco bill, referred to as House Bill 30, that had been sitting moribund for 18 months and with the help of Verizon, slipped a provision in that would have stopped the Philly project in its tracks. The provision forbid any city in the state from building it's own broadband network for which a fee is charged. It was approved in the 11th hour of the lame duck legislative session and sent it to the state senate which passed it with no public notice.

The action caught everyone by surprise, but most dismaying was that the bill would be on the governor's desk for approval before the end of the month. The city only had a week or so to reply. In swift testimony to the public's support for a concept still in the planning stage, several thousand e-mails, phone calls and letters were unleashed on the governor's office. Several public policy and media reform groups, including Common Cause Pennsylvania, Prometheus Radio Project, Media Tank and Penn PIRG, led the statewide grassroots efforts to encourage a veto.

In the aftermath, the bill passed but Philadelphia won a critical exemption from the law and a guarantee from Verizon that they wouldn't interfere with the city's WiFi efforts. While this particular victory was sweet, the political battles in Philadelphia and many parts of the country have not ended. When asked if there was a way around some of these, Dianah replied "I haven't found anything that's unavoidable. A lot of this has to do with where we are as the leader of the pact. You're out front testing new ground. But every community needs a telco and cable strategy. Always monitor the politics surrounding municipal broadband."

The political issues are going to be different depending on where you are. In some cases you won't be able to get telecos interested in bringing broadband to your area, so they may not be a problem. In some cases, they'll fight you at the state legislature and the incumbents are already fighting it in Congress. How are you going to educate your elected officials on the benefit of the technology? Whether you have a strong mayor, a strong council or county commissioners, you have to work closely with them to get them on board because their support is critical in these political battles.

Achieving victory on the political battlefield

If you must do battle, understand that your two opponents are 1) the incumbents fighting what they fear as a critical threat to the Achilles heel of their business, and 2) a political philosophy that enshrines the concept of free market forces above almost everything else. These require different strategies.

The incumbents are a problem because fear makes conglomerates difficult to reason with and forces you into these life-or-death struggles such as House Bill 30. Philosophically-driven politicians aren't much fun either, but at least cities have a philosophical counter punch that can be quite effective.

The strength of incumbents in their legislative counteroffensive to muni WiFi centers around money and knowledge since politicians always want money and never have time to learn all the details about everything they need to know. "We certainly see the 'money' factor in the 14 state legislatures that are trying to stop municipal broadband," states Jeff Arnold. "The political reality is that companies that provision broadband in other ways have spent a lot of money with politicians in terms of campaign contributions. We saw that clearly in PA. That's what that was about. A whole lot of campaign contributions."

The money issue is why I recommended earlier that you first deploy municipal broadband as a government business operations tool, and then build a strong support base of business people. A few thousand angry citizens can produce results, but sometimes it's easier for several mayors in the state to rally a few dozen key business people to apply heavy pressure as needed. For some reason, there are legislators who seem to respond a little better to a phone call from a millionaire or two than they do from the average citizen. In the ideal world, having a vocal coalition of community activists and businesses laying on the pressure is best. So, how do you rally the business community around muni WiFi? Point to three hot-button issues.

Bringing businesses into the legislative fray

Unless a company has the clout to whittle an exception from city council, most businesses pay city taxes and fees. Point out that any smart business wants those tax dollars maximized to get the greatest efficiency in services that the city delivers? Everything from building inspections to paperwork processing and traffic control around their places of work will improve with municipal broadband.

Second, public safety issues are business issues. What would happen if business owners' plant, office buildings or homes suffered preventable losses because their city is hindered from improving its communication infrastructure and operating procedures? As hurricanes Katrina and Rita painfully illustrated, wireless text messaging might be the only communication lifeline business and employees have after a natural or other disaster hits.

Third, most responsible business managers would rather spend $20/month for high speed access for each mobile worker rather than $60 - $80/month for slower speeds. Cities driving partnerships in which a new breed of ISP leverages the latest technology to deliver a superior service at a better price gives businesses a better business technology option that directly impacts their bottom line.

Knowledge distribution is power

With technology-related bills, most local or national elected representatives who don't understand the features and business or social impact of the technology in question rely on someone else to give them that information. Typically, industry lobbyists are the ones who roll in the font of knowledge from which these politicos partake, which needless to say isn't going to paint a pretty picture for municipal broadband.

You need an education campaign specifically targeted to your representatives that not only presents key arguments in favor of muni WiFi, but also delivers the same types of technology-made-simple documents recommended for use in your community relations efforts. Control the discussion and impact bill-writing in your state by influencing what representatives learn about wireless. How you execute this education campaign is dictated by the local circumstances and personalities of your respective legislative bodies. But it is imperative for you to run such a campaign.

Even if your state legislature isn't threatening to prohibit muni WiFi, you have to worry about other laws that can adversely affect your initiative. For example, in Westchester County, NY, an effort took flight to pass a bill requiring Internet cafes and commercial businesses that use wireless networks to take basic security precautions to protect private customer information from potential data thieves and hackers. Fines will be assessed for those that don't comply. Gauging from the resulting blogs written by very tech-savvy people, the intent is commendable, but the approach that the county is trying to codify might be flawed.

The way to minimize these potential minefields is to build a rapport early on between your project team and two or three representatives who are likely to be the main people driving technology bills within a particular body. Newman observes that it's usually the person most knowledgeable about technology whom others rely on to shape how they vote. Having a "trusted advisor" relationship with these individuals keeps you in the loop so you can be proactive rather than reactive while enabling you to gain some level of counterbalance to lobbyists' activities. You want a seat at the table when these late-night bills are being drafted.

Rizwan Khaliq, Global Business Leader for IBM Digital Communities points out where knowledge holds the power to remove the contentiousness from the relationship between municipalities and the incumbents. "There were a lot of sound bites put out early that generated a knee-jerk reaction from certain service providers. I think that as the business models and the whole thought process mature, government as well as service providers will look for a way to jointly come to the party rather than kicking each other out of the party. They can mutually benefit and even help each other evolve the model so they leverage the capabilities of broadband."

Knowing where those points of mutual interests are may be a little difficult if the incumbents persist in such "scorched earth" legislative and PR tactics we've witnessed so far. Perhaps what's needed is a third participant at the party that has a comprehensive view of how all the parties can benefit. Consider some of the large systems integrators to play this role in bringing government and incumbents together.

When the opponent is philosophy

In the philosophical battle you have conservative state legislators, just as you do in Congress, who say that cities should never be competing with the private sector. The incumbent telcos play this card a lot. You need to bring out your trump card. Ken Fellman, Mayor of Arvada, Colorado and former Chair of the National League of Cities' Information & Technology Committee, spells it out this way.

"The state or Federal government shouldn't be telling local governments what to do. It's easy for us in local governments to argue this and a difficult thing for conservative legislators to argue against because, while they like to be pro-business, most conservative philosophy argues for local control. Whether it's a town of 500 people, or a city of millions, if the citizens examine the issue and they say 'we are willing to spend our tax dollars, we're willing to vote for a tax increase or issue bonds because we think municipal broadband is a good idea in our little town,' then who are you at the state capital to tell us we can't do that? We thought you supported local control? Isn't that what you're party's all about? This argument resonates."

The decision to deploy municipal WiFi is no different than when that same town decides they should issue a bond to build a new road or new baseball fields. Or they want to spend money for bigger pipes in their water system to get more capacity to address the increase in population. That's a utility, right? What legislature is going to tell that local government they can't build it. Then why isn't municipal broadband the same thing? In Colorado, an anti-muni WiFi bill started off with very restrictive rules. By using these arguments, Ken and his colleagues got it watered down to the point where local communities removed the threat.

"In order to fight this battle, look for coalitions because this battle isn't all liberal Democrats versus conservative Republicans. When you get into these debates, a lot of times it's the rural areas that aren't covered and they're the ones who will benefit the most, at least initially, by putting in a wireless system. A lot of those areas are represented by conservatives who may likely look at the debate and say, 'this isn't just business versus government, it's about local control.'"

Carry this approach up to the U.S. Congress as well. Align Senators and Representatives from rural areas with those whose states that have large metropolitan areas facing significant social and economic issues. You can count on organizations such as the League of Cities and the National Association of Counties to keep these efforts in play as 2006 likely is the year when this issue will peak in the Capital.

At the local level is where the philosophical battle can really hamstring your initiative. When the incumbents come in and say it's unfair competition, you have some fairly conservative communities that respond favorably to this. There are folks you'll never be able to convince that any kind of municipal service is ever going to be a good idea because "when the market decides it's ready, it'll get here." You have other citizens who think that this is an appropriate role for local government. Success, therefore, is determined by how well you marshal pro-muni WiFi forces to work for your side.

The same wholes true at the county level where some of the commissioners are resistant because of their belief in the private sector. Jeff offers a tact that enables you to turn the gods of market forces in your favor. "In most cases, the citizens want what they want when they want it. People are beginning to understand wireless more and more. Broadband acceptance is moving quickly because anyone who's ever experienced it doesn't want to go back. There's an interest among those underserved segments to get any kind of broadband. If this is true in your county, discuss the market demand and show how municipal wireless makes sense because it's the cheapest and fastest alternative."

Once you build support based on market demand, you've elevated the discussion so that people with different philosophies can come to the table and talk about how to get more broadband and more competition into your communities. In some cases that will be the city itself providing services, in others it will be the government seeking private sector partners. You may find that the government/business partnership option is what enables you to find middle ground with the private sector advocates, and you can move forward.

What if the decision is fait accompli?

In those states where legislation is already on the books prohibiting municipalities from selling any type of broadband service, it's probably safe to say that consensus building is definitely an uphill battle. One option I recommend is go back to Chapter 2, build your business case for using the technology to improve government communication and operations, fund the network and then provide the extra access capacity as a free service. Make sure the system allows you, as does the one in Corpus Christi, to sufficiently secure the city's network operations, servers and other resources from the general public's activities.

Obviously, if you do this, add disclaimers that no customer service and support will be provided, everyone is responsible for their own security and enhancements are possible once their legislators see the light and reverse the ban. While this network isn't going to be as effective and beneficial for your entire city as Philadelphia's network, you should see many grass roots efforts blossom up to provide support and constituent programs in the neighborhoods. Your social and economic development efforts will still have a platform as long as there aren't any fees charged. Companies' IT people should be able to tap in and provide some basic mobile application integration with the network. In fact, a little eco system of consultants might spring up to smooth things out.

The end goal here, by the way, is not to create a permanent workaround. That's just a short-term benefit. What you really want to do is get the laws changed. Once you get people addicted to broadband, they're going to want to do more and they're going to want the next evolution of the technology, whatever that may be. As the masses start to agitate, inform them that the road to their desires is blocked at the state capital by the representatives of the people. Then turn your citizens loose. You read about House Bill 30 and what people can do in just a week when they get sufficiently motivated. Sometimes you have to play a little hardball to get the consensus you want.

To sum it up

Consensus building is the lifeblood of a successful municipal broadband deployment. Plan on it, insist on it, execute on it. Ken concludes, "I believe, and others I know believe that good local leaders try to see the future and where we need to go. Then we help people get there a little faster than they would if you weren't pushing them. If you're really good at it, you have people think they got there all by themselves."

Chapter 6

Understanding the Needs and the Will of "We the People"

During the country's 1976 bicentennial celebration, caps and t-shirts bearing the words "We the People" were quite popular in Philadelphia. As Executive Committee member Robert Bright said in the last chapter, this phrase tends to sum up the mindset of the city's citizens when they rally round certain causes. Whether or not these citizens feel their local government accedes enough to this philosophy when launching major initiatives, they should be pleased with the significant efforts Wireless Philadelphia made to understand what the people want from muni WiFi

Two things were evident as I reviewed details of Philadelphia's use of focus groups and town meetings. First, these were incredibly effective tools for building consensus because their impact went beyond helping the Executive Committee re-think some of its assumptions while formulating its business plan. Second, this was a better effort at understanding end user needs than even major corporations with large budgets do when it comes to wireless applications.

As I read officials' statements in the stream of press releases coming from cities and counties gearing up to explore WiFi initiatives, I sometimes wonder how much due diligence their staffs plan to do before moving too far down the path. And how extensive will those efforts be? In this chapter I present an in-depth look at Philadelphia's focus groups, their impact on the planning process, and discuss the role they played beyond gathering feedback. This process is the standard for due diligence that governments should try to match or exceed to be sure that what you ultimately deliver meets the desires and needs of We, the People.

Philadelphia's atypical focus group efforts produce great feedback

One shortcoming I've seen with wireless deployments over the past four years is that organizations don't spend enough time listening to the people who actually use the technology. It's not necessarily a function of money. Companies with lots of money are just as guilty as cash constrained groups. Sometimes I gather it's the ego of the people driving the process and the assumption that they know best what users need. Other times project managers are heavily time constrained because of so many other commitments. However, the more end users you listen to, the better the application is that you design and deploy.

You face a significant challenge trying to gather meaningful feedback about muni WiFi is because this is new technology, many people don't understand it, there aren't a lot of deployments to reference and the potential end user base can be in the hundreds of thousands. Assessing the wireless needs of a municipal work force is reasonably manageable. But to pursue a grand vision of resolving social needs for a diverse population, spurring major economic development and making a town or city more desirable for visitors demands an aggressive needs assessment effort.

It is possible to do quantitative research in which you mail out lots of surveys and conduct numerous phone or on-the-street interviews. This works when you need general feedback about easy to understand products, services or concepts, and you want a snapshot of what citizens think. General public meetings are good for this as well. However, for complex products or concepts and new technology, you need qualitative research where people in the process have time to get their minds around what you're talking about. Wireless broadband definitely requires the latter research option.

Philly sets itself up for success

In order for the Executive Committee to meet its aggressive three-month deadline for developing the business plan, they pursued several tasks in tandem. Area universities, including Temple, were contacted and student/professor teams were recruited to gather industry and financial data to help the committee make choices regarding technologies and business models. Community business leaders and representatives from the school district were contacted to assess their respective perspectives on municipal broadband. And the Committee decided to use focus groups and town meetings to capture feedback from a broad spectrum of stakeholders.

To develop the focus group process, the Committee brought in Julie Fesenmaier, Associate Research Director at the Cochran Research Center in Temple University's Fox School of Business and Management. The Ronin Group was chosen to facilitate the focus groups and the town meetings.

Focus groups are different than town meetings in that the focus groups have fewer participants, maybe 15 – 20, and are conducted as an intense roundtable discussion in a closed setting. Philly held 20 and each group was comprised of people from a particular stakeholder constituency. A city this size should have anywhere from 20 to 30 sessions to make sure you get the broad representation that's needed if you want to an accurate picture of what your municipality thinks. A town meeting is a larger gathering, participation is open to all communities, and these may attract 50 – 60 people. While the discussion is still controlled, it is not as tight and intimate as a focus group.

An important milestone was to define who the stakeholders for the focus groups should be. The tendency is to just suggest technology users. But Julie strongly advocated for the Committee to gather feedback from representatives of all the stakeholders, whether their comments were for, against or neutral. This included inviting the corporate sector that's not supporting this, though that group wasn't too willing to share any opinions even though the Committee tried to set up a focus group for them. Julie says that "if I hadn't pushed the Committee on this, the quality and accuracy of the feedback would not have been nearly as good. But they let me define who the stakeholders were going to be and what kind of questions we were going to ask them."

Identifying key issues

Julie created an eight-page workbook for identifying key issues, and the Committee completed this together in a session. It was a vision exercise in which everyone had to articulate what they saw as the best case scenario for the network and what was the worst case scenario. It's a random thought process to see what comes out. "All of these diverse people, brought in from education, health care, general businesses and so on, had to buy into a multi-faceted definition of what this initiative meant. In the session they also had to list categories of stakeholders the focus groups would represent."

The workbook started with questions such as "What benefits does a wireless community bring to its residents?" and "What does that perfect wireless community have?" Some of the leading responses from members were "accessibility," that the network not be limited to time or place, that there be unlimited information, and "opportunities," the network has to offer equal access, business opportunities and education enhancement.

Members were asked to define the characteristics of a wireless community that has good services and infrastructure in place, and a wireless community with poor infrastructure. They cited cost effectiveness, reliability and security among the positive characteristics, and bad service, bad publicity and lack of user friendliness among the negatives.

The Committee had to describe the main functions they saw the network serving in the community. Leading responses included "increasing connectivity between businesses and customers" and "improving the quality of living." There were also questions about how to ensure the sustainability of the network, to which some of the key factors were continuous funding and financial viability, partnerships and collaboration.

Finally, the Committee worked through a list of 30 values in order to assess "What values drive the development of this community technology program?" Looking at each value, what's of great importance or less importance? The top values for the Committee were being innovative, providing leadership, being reliable and being responsive. Julie wanted to be sure that these values were reflected in what the focus groups discussed. At the end of the session they developed a list of participants and questions for the focus groups.

Facilitating the focus groups

Though there are firms that coordinate all of the logistics of focus groups for a client, the Committee decided to do things differently," states Ronin Group Principal Robert McNeil. "Various members of the Committee had ideas and contacts for potential participants. We recommend that, if a city is going to select participants, look at the stakeholders groups they know and get as broad an audience as possible of people who have an interest in wireless."

Committee members reached out to several colleges, hospitals and chambers of commerce. They went to individuals who represented large groups. They also did the cut economically, soliciting different organizations and community groups within neighborhoods. This was easy for Philadelphia because it's a city of neighborhoods with very active organizations such as the Asian American Chamber of Commerce.

The Committee also wanted people who didn't belong to an organization, but who were recognized members of particular communities so there would be representation of a geographical area rather than from a particular group.

You can ensure diversity of opinion even within a group of seemingly similar people. Participants can be a from a group of colleges, so they're homogenous that way, but if they don't know each other then they're still diverse in terms of how they might have used wireless technology or may want to use it. You will get better information than if you bring in a fraternity group or some other group in which people know each other.

Running focus groups effectively

Needless to say, asking the right questions makes all the difference in the world to the outcome of the focus group process. Some of the Committee's questions focused on pricing and how much would people pay (one focus group participant replied "$1 less than my AOL connection"). There were questions about what people thought the benefits were. A major question was did participants think that the city should operate the network along the lines of a pubic utility, or should it be run by an independent consortium or a cable company.

A lot of energy went into the question design and refinement process that is atypical in standard focus groups. Julie and Robert took the Committee's questions, added their own, and then went back and forth many times refining questions until they had 25 they wanted to send to the Committee for final approval. They had meetings after the initial focus groups to talk about the questions some more. "This question is getting flat, what do you think about changing it around?" About six didn't really produce that much information, so these were replaced.

The classic focus group format is to have people come in and you ask them a set number of questions, usually in sequence going around the room to get everyone's answer. But when you do it this way, people start influencing each other's answers and you don't get the type of candid responses as you might in other formats. Robert's approach was to ask participants to write their responses first on index cards, then get them to state their answers.

"This took a little bit longer, but we felt we got better information. If people work on something individually and then they talk about it in a group, they're more likely to state their own opinion than that of the person before them. By writing it down they formulate their opinion first." Another thing Robert did to facilitate discussions was follow the energy in a meeting. "If it was getting hot about something, we could pursue that. For example, it got hot around the digital divide in a lot of sessions."

Another thing you want to do is control for over-speak so you hear from everyone, not just the assertive people. Robert intervened pretty quickly if someone started to dominate the discussion. "We'll say 'we really heard from you on that question, let's hear from some other people.' You want to create a 'protected' atmosphere to ensure that each person's opinions is fairly independent, which happens when you give people a sense of freedom of response and they feel like they can speak candidly."

Encouraging candidness is important if you want to maximize the main benefit of focus groups, which is the quality of responses you get when you can delve deep into people's thinking. You get to ask follow up questions, or have interplay going across the group so someone can add to what someone else said. This way, you get a smart group that responds. You're not dumbing down the process by asking all the groups the same questions. Also, when people from diverse groups push along different paths, but come back with some similar conclusions, you know you're coming to some collective consensus that will help unite the general population.

Philly goes the extra mile to maximize benefits

To get the most benefit from your focus groups, you need more than a passive involvement with their execution. Don't let things rest just with the facilitator. The Committee received reports after each session and a video tape to review if they wanted to get a better sense of some of the emotions in the group, and members also were invited to participate in the sessions.

Those who attended were able to push the group even more than the facilitator, though he still maintained control and direction for the sessions. Robert recalls a group of African American businesspeople in one session where the question came up about whether wireless would be useful for low income residents and how. Participants were answering when someone from the Committee said "you're not really addressing the question of will there be a digital divide if we bring in wireless and everyone can't access it."

A person in the focus group shot back, "where have you been? There already is one! Some of us have computers, some of us don't. Some of us don't even have a clue. The digital divide is all around. It's not that this is going to increase it or decrease it, it's already here. If you're going to address it seriously, then you're also going to have to provide access to a lot of folks. Access might actually mean training, it might mean computers, or it could be some other kind of solution that you may not be thinking about." The Committee got a lot more information at a level where they could feel the intensity of the disenfranchised people in Philadelphia.

Besides good feedback, you also want to convert focus group participants into muni WiFi champions. What the Committee did was to create the vision of a wireless Philadelphia and gave people an opportunity to contribute to the vision. What's more, they told participants that a small part of the network was already up and people could go to a section of city and try it out. "This is really important," Robert says, "because it wasn't an abstract concept at that point. If anyone wanted to see how it would work they could go down by the LOVE statue across from City Hall and see it. A number of people had actually tried it before the focus groups.

It was also beneficial that Varinia and her group created a brochure that showed this section of coverage. In the meeting they could hand these out and tell everyone to try it out. Her team was quite savvy about focus groups being a marketing force. The participants were movers and shakers, the connectors. They would go back to their constituents with the brochures and say "look, this is what's happening. I was just in a focus group and these are the things we can do with wireless." Robert asked participants to invite people to the town meetings. Participants were sales people as well as believers when they left the sessions.

Making participants feel special is another way to maximize the benefits of the sessions. One of the challenges running focus groups is making sure that people attend. It's a waste of money and time to have a facilitator, a nice room but only three or four attendees. A subgroup of the Committee did a really good job of providing that added personal touch. The staff contacted participants to make sure they were coming. Everything was top shelf. Food was in place when everyone arrived. The sessions were invitation only and they were in nice rooms with great facilitator equipment. It was clear to attendees that the city was very serious about this from the very moment it started.

Robert believes that all of these little details "show that a city is welcoming, appreciative that people gave up a half day to make an important contribution and interested in what participants have to say whether you agree with their opinions or not. What was nice in Philly was that we had high-level doctors from prestigious hospitals and community leaders from some of the low income sections of the city, and everyone was treated the same way. People felt a little special for giving their input."

Town meetings bring it all home

Philadelphia has a history of town meetings. For example, if the regional public transportation system (SEPTA) is considering a fare increase, there's always a big town meeting about it. The city convenes these for many issues about which they want feedback. These two for municipal wireless were broadcasted on one of the local TV networks so people could watch and also link in via phone. "The dynamic that's different from focus groups is that you don't know who's going to show up," Robert remarks. "Who usually shows up are the interested parties, of course, so we expected representatives from general communities, business people and the telcos." Well, two out of three isn't bad. The telecom and cable companies didn't show, or at least didn't make their presence known.

The meetings produced very engaging conversation going back and forth that flowed mostly on its own. Robert just opened up the discussion after Dianah's remarks and fielded questions so people felt there was an organized process. It went very smartly. People asked questions and others gave strong opinions about what they thought. Both sessions were very favorable for broadband. There was caution by some people asking were they going to get this at the cost of other services being cut.

"From my experience with other town meetings, these two were well attended and we got a diverse audience," according to Robert's critique. "I was expecting fewer people or more people strongly focused on one side of the discussion. We didn't have people who were advocating either for or against this. The meetings were more like information-gathering sessions. I was surprised because a number of people actually used it as a focus group and brain storming process. 'Well, if we do this, do you think we could use it for that?' There was the fellow speaking on behalf of seniors asking if they would have access to helpful information about health conditions. Most of the time you have governmental meetings that are more adversarial."

The entire Executive Committee was at both meetings getting into discussions directly with participants. They uncovered from attendees a real desire to be part of something that sounded way cool to people. The concept of a wireless Philadelphia pretty much caught fire, which was evident by the questions. "We can be part of this? We can really make this happen?" Once you get so many people from all walks of life taking time from their schedules, volunteering ideas in a public forum, and building on each other's ideas, you're another big step closer to matching the right wireless technology to constituents' needs.

Though you can do either focus groups or the town meetings first, Robert thinks it was good the way Philadelphia did it. "We became good at asking the right questions in focus groups, so we were really up on what the themes and major issues were. When we got to the town meetings I had a thorough understanding of what the Committee wanted to know, and it was rather simple to follow up on attendees' questions. When someone presents a question, it's good when the facilitator can ask 'well are you also talking about...' and probe what it is that people want to know. Since the Committee had the results of focus groups, they too were knowledgeable and better prepared for the larger meeting."

In the final analysis

The tone that you set for the focus groups has a lot of impact on what you get out of them. Philadelphia tapped into what Robert calls a "creation orientation" instead of a "problem-solving orientation." As a marketing person who's worked on both types of projects, I have to say that you really get more mileage with the former than the latter.

When you try to create something you bring something new into being, and there's a lot of energy you can get around "wouldn't it be cool if...?" brainstorming. Or President John Kennedy's approach in the 60's of presenting the vision of going to the moon in 10 years and challenging those around him to create the best way to make it happen. You get this incredible vision out there with lots of people contributing to it because they can be a part of the dream. The problem solving orientation is typical when people deal with government. Instead of trying to bring something new into being, you're trying to make something go away. "Make my taxes go away." Make this or that problem go away." There's often not a lot of positive energy in these types of meetings.

"So we weren't trying to find out 'what are all the problems we can expect with putting up wireless in Philadelphia,'" remarks Robert. "Because of the creative orientation, there were many ideas bubbling up in all of the focus groups and town meetings. 'I can be sitting on the bus and contact a restaurant to make reservations for dinner using my handheld?' 'You mean I can use that thing to Google a company while I'm waiting in the lobby to meet with them?' People were building on each other's ideas. That was part of Dianah's thing. We were going to make this network come into being. And believe it, this is a vision. Yet it's concrete enough that people actually believe it can happen."

If the members didn't get an insight to the will of the people, at least they got an understanding of the desire of the people. The Committee was making plans for what they wanted and what they thought the people needed, yet sometimes they were surprised by the responses of the focus groups. They really needed to see up close and personal what it is that people are really interested in doing with the technology. They were able to come back after a focus group and say they had better insight to the emotional tone.

Knowing what people want and the emotional intensity of their need is important to the discussion about political consensus building. Robert observes that "there were people who didn't think muni WiFi was worth while at all. Even though they weren't in the majority in the focus groups, they were still there and they were very candid. Politicians need to see all of these viewpoints on the discussion because sometimes they are not in sync with what people feel."

Julie had a student go to all of the sessions to take notes on specific observations. His and Robert's reports were all pretty extensive. After reviewing the documents and some of the tapes, she wrote the final analysis report. It takes a fair amount of work after transcribing everything to make sense of what people said, and the researcher has to make sure that their personal leanings don't filter into the results. This is why the tapes are important.

If you look at any marketing campaign that's based on focus group, they're rarely are as extensive as this. "So I feel pretty confident of the results" Julie concludes. "I believe that the users determine the way these focus group go and the results you get reflect their true feelings. I would like to have had the media there, but we didn't think about them until it was too late. The media comes in with different principles and values about communication and I'm sure their opinions are not unanimous. And it could have been good to hear what the Comcasts had to say. I talked to some of them informally, but they just said 'it's never going to go,' and brushed everything off."

Focus groups – the view from within

Having explored the key issues of running effective focus groups and their role from the perspective of facilitators and organizers, I want to give a little insight from two participants in the sessions. Betty Mon is CEO and Senior Strategist for Mon & Associates, a Center City Philadelphia business and market development consulting firm that focuses on providing innovative services to help business in the Asian

American community expand. Brad Tabaac is owner of Friendly Pharmacy, a local business in the Norris Square neighborhood for many years.

Here are some of their thoughts on their focus group experience and on the Philadelphia wireless initiative. My interviews with them were before Earthlink was the announced winner of the bid for the project.

1. Was there a value for you in participating in the focus groups?

Mon: The value to me was my ability to contribute, to have a voice in the project. As a newcomer to the city, and based on my experience in government, the private sector and as a Master's law student, all of these together gave me a unique perspective compared with some of the folks in the focus group who are in one sector or another. I don't know if I learned something other than gathering knowledge about wireless. I was happy that Philly was undertaking something like this and being one of the forerunners in technology.

Tabaac: Only in that it was a focus group co-sponsored by Norris Square Civic Association, which is the neighborhood where the pharmacy is located. They asked for people who could give thought to some of the issues that were on the table. I participate in the community, so from my standpoint I was just giving back to the community I serve. As a merchant I felt I could add something to the conversation.

2. When you first heard about the potential benefits for businesses that the Philly wireless project would make possible, what were your initial thoughts?

Mon: I'm a big networker here and in my dealings with people they've identified me as someone who knows what's going on around town. So a friend asked me to participate in a focus group. One of my initial thoughts was, I understand why we need to step forward and do something, but why are we allocating all of these resources to handle wireless when we have so many other problems going on. Philadelphia, because of its history, is important to the nation as a city. Yet it has a lot of issues that it's dealing with. This was discussed at length in the focus group that I was in. How does a city government allocate its resources so that the city is moving forward, but also take care of all the other needs?

Tabaac: My first reaction was that many businesses not on the Internet would be able to get access that enables them to operate their businesses more efficiently.

As an example, my staff is asked a lot of questions that need to be answered by accessing reference guides. A doctor may ask about the ingredients in a rare product, so I Google the item. I'm able to give the person the right information in seconds because I have access to a T1 line. But there are a lot of small companies only hanging on by a shoestring that aren't going to have an opportunity to expand, in part, because of the their inability to get onto the Net at fast enough speeds.

On the community side of things, I'm sure there are people with entry-level computers who can't afford AOL. So their kids aren't able to get on the Net at home. They fall behind the kids who are on the Net. Today kids as young as seven or eight have access. If the city becomes wireless, more of them will have opportunities to get online.

3. What should the role of city governments be in bringing this kind of technology to their constituents?

Mon: Who's the proper entity is a big question. I thought a quasi private entity would be the most successful, with city government being part of the steering committee. Philly is at the helm of this, that's ok. But there were reservations about the ability of the City to really make this a successful program. Who's going to make it happen and how much is it going to cost? If a business is going to run it, then which company? Is it through a partnership with the city government? I don't know what the answer is. However, I believe this project should move forward.

Tabaac: That's a tough question. I understand that Philly doesn't want people left behind from a technological standpoint, so it's to the city's advantage to foot the bill. This is one of the things we discussed in the focus group that I didn't have a feeling about one way or another. People were saying 'the service should be free, it should be free.' I'm saying to myself, nothing's free. It has to be paid for by somebody somewhere – whether everyone's taxes go up $2 or whatever.

I don't know where the government should start, or where they should stop and then let the people become responsible. If putting up this wireless world is not that great of an expense, then maybe it is something that the government should provide. I don't know if it's the government's role to bring people in to tell constituents how to use the technology, though. But if the government is willing to put up the infrastructure, then maybe they should charge some private community-level peers to get involved. Norris Square can get a newsletter together to say 'come in and we'll have a presentation and show you exactly what you can do with the technology.'

4. Do you see municipal WiFi changing the way your business or other businesses operate?

Mon: I can definitely see that for mid- to large-size companies doing business across state or county lines and for people who do a lot of traveling, the mobile aspect of this technology will affect them, though it may not affect me, per se. Most of the small companies that I deal with now are retail or local, so they always have hard wire connections. They may use cell phones or their PDAs to get them through the day. I can't see any of my clients using wireless to any great extent for mobility purposes at this point.

I do like the aspect of the service where I'm just paying one fee to access the Internet. Right now, if I'm home I have to pay, if I'm in the office I have to pay for that connection, and if I'm traveling outside that's another fee. That three access fees and it just doesn't make sense. For a small business with limited resources, which one are they going to pick? It will be for the location where they spend the most time. If wireless is covering a large area where business owners live and work, then it makes sense for them to buy into the network.

However, whether people buy into the service or not is ultimately a marketing issue. How comfortable am I going to feel about wireless broadband? With regular broadband I know what I'm getting. If I'm being offered $15 dollars a month for wireless, what am I getting and why is it so much cheaper? There has to be some educational effort so that I'm comfortable with it and my clients feel like they can rely on the service.

Tabaac: Not really because I'm already wired. Would I save money if I got rid of my T1 line? Sure, but having wireless won't change much of what we do currently in terms of operations. In my office I have five workstations on my LAN, three have Internet access. My hub is maxed out. If I bring my laptop in from home and want to get on the Net without having to take a computer off the LAN, I have to invest in hardware and additional wiring which I wouldn't have to do if the city has WiFi in a protected way.

Now let's say I want to expand my service and build a new patient counseling area. With the City's WiFi I could throw more workstations on my LAN with no extra set up expense so if a patient has questions I just jump on the Internet to get their answers. Or if I look at my delivery set up right now, it's all manual, so wireless probably will help me here. There is technology out there that, similar to the UPS person, lets any driver use a digital pad to get customers' signatures and then upload information from their trucks to the distribution point so they can better track packages. If my drivers are in areas that are wireless, they could use this technology to communicate with me via the Internet.

5. Do you see businesses reconciling their needs with those of the City?

Mon: The City sees itself giving small businesses less expensive access. Is this aligned with small business needs? I believe it is, but I don't think small businesses are conscious of the issues. They don't take time to ask 'is broadband really good?' They're so overwhelmed with everything else they're dealing with that, when they figure out they need access, they'll just reach out and buy the best option they can find. Later, when the network becomes a matter of fact, businesses will take it for granted like any other city service.

In the focus group we talked about other issues, though. There were a lot of "why" questions. We had a big discussion about costs. What is reasonable? In the preliminary stages, who's going to bear this? If it's the business community, what kind of marketing efforts does the city have to take to bring them along? You have local businesses, transient business folks, students and your local residents. At what point do you charge them? People seemed less concerned about defining their needs than knowing how much they'll have to pay.

Tabaac: I don't know what the city's needs are besides just being able to say we built this system. But why the Mayor does it doesn't bother me as long as what he's doing has merit. As far as small merchants, if someone gives you something that you don't know what to do with, it's worthless. I know businesses that aren't spending $15 a month to keep the sidewalks clean. So these guys aren't going to want to spend money for something that they don't even realize can help them. It seems like the gating factor in whether WiFi meets their needs is how much they're willing to learn about it on their own and how much someone is willing to teach them.

Can RFIs (requests for information) subvert the value of focus groups?

Bringing all of your constituents together to build creative energy and enthusiastic support for your municipal wireless initiative sets the stage for the important next step of the implementation process, matching the right technology with user needs. In these early days of municipal broadband, I feel the use of Requests for Information (RFIs) that's occurring in some cities presents a danger to this next phase if it's not timed and managed properly.

The proper role of the RFI

RFIs are issued by organizations that realize that their staff doesn't have a lot of expertise in a particular technology and its use. They want to bring a number of experts together to brief the staff on the basics of the technology. The information and ideas that come from this process help the organization more effectively assess its options and develop a better RFP. The logic is sound, but there are a couple of potential problems.

For one thing, you may be eliminating input from the vendors who can do you the most good. Some RFIs are more than a request for a sit-down roundtable discussion and technology briefing. I read through one city's RFI and compared it to Wireless Philadelphia's RFP. The RFI asked for essentially the same level of details on how a vendor would deliver muni WiFi that the RFP solicits. However, this rather open-ended RFI lacked specific guidelines or requirements in many areas on what the city expected from vendors and what was needed by citizens, which is what you find in the RFPs of cities such as Philadelphia and Portland.

To give you an example of the situation, one line item in the RFI required that WiFi 802.11b/g technology should be a component of the application. That's pretty much it. There were no specifics given. In Philly's RFP, there's also a request for 802.11b/g technology, and it must include features such as:

- Compliance to IP56/NEMA4 dust and water ingress ratings for all outdoor-mounted equipment.

- Support for ambient temperature ranges of –40 C to +50 C for all outdoor-mounted equipment.

What you have with the first document, from a vendor's perspective, is either an overly broad Request for Information or a poorly written Request for Proposal. Both you and vendors are in danger of getting burned.

As Paul Butcher from Intel observes, "A good response to an RFP requires hundreds of thousands of dollars in time and effort. Even though an RFI is supposed to be the city asking vendors for some general ideas, it's still going to cost a lot of money to come up with a written document that may be 100 pages long. I mean, how else do you formally respond to something like this as a vendor [referring to the RFI]? You put one or two of your knowledgeable people on the project, or maybe even a whole team, who by default are very valuable to your company. They spend days putting their thoughts together to share information for free just for something that may not even work out. And there's still the eventual RFP process."

Even large vendors and system integrators are leery of broad RFIs. Should they respond favorably to the RFI, there's no guarantee they'll have an advantage when the RFP comes out. In fact, a city could significantly change some requirements based on information gathered through the RFI, requiring yet more work. With small or mid-size cities, there's less incentive for a vendor to jump through these double hoops because the size of the eventual contract is smaller. As a result, you miss out on getting valuable ideas and approaches that can come from smaller vendors because they can't cost-justify responding to your RFI.

What if there's no will?

The second potential danger comes when the overly broad RFI doesn't reflect the level of in-the-trenches research Philadelphia employed. This puts the cart before the horse. How can a city possibly undertake a massive project such as this to address the will of the people if you have only a basic-level understanding of what the people think or want? This leads to is the great-idea-in-a-bubble syndrome.

A lot has been written about Google's response to San Francisco's RFI with an offer to provide free WiFi services, with their investment presumably offset by ads running on your screen when you log into the service. Free WiFi? Great idea. Has anyone convened a hundred citizens representing a cross section of San Francisco's constituents to ask: 1) would they use a network that hits them with ads and special promotions based on the location where they access the service; and 2) what is the likelihood they would respond to these ads?

If many people answer the first question "no," that would seem to offset the City's goal of enabling universal access. If many answer "no" to the second, that would appear to bode ill for longevity of the partnership with Google. Of course, the answer to both questions could easily be "yes," in which case everyone is happy. The point is, if you don't ask these and other questions before issuing an RFI, you risk the success of the project.

Some cities may think that vendors with experience in deploying muni WiFi networks are able to respond to an RFI with the appropriate recommendations because they know what they're doing. It's possible. But since muni WiFi is so new and every city is different, that's a pretty serious leap of faith. Many of us are still stumbling around in the pre-dawn's dim light and the new day's sun of comprehension hasn't fully broken the horizon yet.

Art of the Possible versus Art of the RFI

Instead of creating an RFI that's essentially a slimmed down RFP, consider St. Cloud, FL's approach, which is an RFI that is neither a major burden on a vendor, nor putting the cart before the horse.

When they started exploring the benefits of municipal wireless, Mayor Glenn Sangiovanni and others went out to California to observe the San Mateo Police Department use of WiFi hotspots. Officers were getting line up pictures sent to their patrol cars, completing reports and performing other tasks that couldn't be done using the speed of cellular data services. What the St. Cloud delegation saw prompted them to probe even further to see what they might be able to accomplish.

"We contracted with HP to have them use their methodology for wireless needs analysis called the Art of the Possible," says Mayor Sangiovanni. "It's a two-day session they did at our facilities in which we brought together business leaders, community leaders and regular citizens. The first day is spent getting everyone up to speed on what the technology is and what it can do. The next day everyone discussed what some of their needs or uses were for this technology. 'How do you see this technology helping you now that you've heard what it can do?' 'Is it of use to you and would you apply it in the business field, or in the public sector?'"

The group generated many good ideas during the brainstorming. There was a follow-up session that brought everyone back together six weeks later after people had time to think about things. The Mayor passed out sticky notes and asked people to start writing ideas which were grouped into categories: Benefit to the private sector; Benefit to the public sector; Benefits to a specific industry. It's overwhelming how many ideas this process produces. It's just a matter of how creative you can be.

Then Mayor Sangiovanni started looking at the technology's impact on the city government's departments. He sent notices to all of the department directors asking the to consider what the technology could do for their specific area, whether it was improving permit processing, inspections, trash pick up or other key operational functions. They were asked to come back with at least one use for this technology so the project team could see the potential cost savings and if they wanted to, move forward.

"This is where we came up with the operating budget impact," says Mayor Sangiovanni. "It was going to save six to eight jobs over the next three years in terms of people we won't have to employ. The cost savings here pays the operational side of maintaining wireless citywide." Rather than ask vendors to suggest business models, St. Cloud came to consensus on their own what made the most sense.

Once a city and its various stakeholders get its technology briefing and do several rounds of brainstorming, it can launch its focus groups if it plans to conduct them. They can pursue a more formal RFI process based on more clearly defined interests in citywide WiFi. Or just simply move to an RFP. Philadelphia's Project Manager Varinia Robinson said they never considered doing an RFI at all. She had consultants and a few vendors doing feasibility studies and a lot of technology testing, so they were getting real-world feedback that was more informative than briefing documents.

Do your own homework

These suggestions for the RFI process aren't about making life easier for vendors, they're about understanding whose role it is to do the homework necessary to determine - and be able to articulate to vendors - the needs of your citizens. This is how you best define what muni WiFi should accomplish so vendors can give you the some valid ideas on what technology is best suited to your situation. As we move past these initial big-city projects that generate headlines and muni WiFi deployments become more commonplace, there are some things vendors won't do anymore. Responding to a labor intensive RFI process is one of them.

Butcher states that "vendors like it when the city has done all of its homework and has committed to focusing on one or two key applications. What are the priorities within government? What do their citizens get from this? Vendors like the idea that a city has tracked where all of its network resources are, such as fiber in the ground. Cities need to know if it has access to key assets such as lamp posts and rooftops that are needed to deploy the infrastructure. It helps vendors design responses that are most appropriate."

Ken Fellman, Arvada, CO's mayor, describes details that will go into their RFI. "Here's what we have – cable modem service from x. We have these business areas, and here's our plan for a business park we plan to develop. Here are the kinds of businesses we're looking to attract. Here's what our education and medium income levels are. What should our cities be looking at. We're trying to get info back from vendors that we can look at it and see what makes sense? Do we have the fastest speeds we need or is there any way to increase that? Are prices such that, if we encourage another competitor, we'll be able to get better service and better choices? From here we'll decide if we need to take some kind of action."

Vendors such as HP, Motorola and Earthlink can give you the technology briefings you need to understand the basics of the various technologies and their capabilities. But the government owes it to taxpayers to do the rest of their homework upfront and figure out what it is that they want. Look at what Philly and Portland did. They took their time and used internal resources to collect all of the information and find out what they needed to know without putting all of this on the backs of vendors.

In Wireless Philadelphia's case, the Executive Committee started their homework with one set of assumptions and a desire to develop a city-owned service. Then they conducted the focus groups while simultaneously expanding their technology research and financial analysis efforts. What if they had produced a broad RFI document instead? How likely is it they would have received or accepted a bid such as Earthlink's? Maybe they just would have gotten distracted, wasted time with an RFI process and then discovered that citizens wanted something else.

Another vendor contribution to the information process that will fall by the wayside is the product give-away. Philadelphia got a lot of mileage from the fact that they could send focus group participants and other citizens to working hotspots that were made possible by different sets of vendors giving the City equipment for each hotspot. The vision of muni WiFi became a concrete, albeit small, reality for those early users. But it also cost each vendor thousands of dollars, plus the staff some of them devoted to the teams running the pilots. It was deemed a worthy crap shoot by vendors who wanted the bragging rights that now go along with deploying the most high profile project in the universe. However, after one or two more major cities start deploying, along with the numerous smaller communities, the bloom's pretty much going to be off of that rose.

Requests for free or borrowed equipment are probably going to be met with what Civitium's Greg Richardson describes as "a polite but firm request to pound sand. The more aggressive vendors might give you a plane ticket to the nearest area where a deployment is already in place. It will be cheaper than giving you the equipment." So, if you want to "make it real" for constituents, you need to wrap this item into your budget for the pilot project.

To sum it up

Now that we've looked at the emphasis that needs to be placed on learning from your constituents what their needs are, it should be clear that this is an element of the implementation process where you don't want to skimp on effort or resources. But don't get so tied up gathering feedback that you don't move the project forward. Dianah Neff always kept the Committee and her project team moving forward and made sure everyone was deadline driven and maintained a sense of urgency.

Smaller towns and cities that plan to concentrate mostly on using the network to enable mobile workforce applications within the government may not be interested in running as many focus groups. However, some level of needs analysis should be conducted with constituents even if you only plan to offer general wireless access. Cities such as Orlando that launched, then later shut down, their networks may have saved a lot of trouble and expense with more qualitative research. Or they might have uncovered needs or interests they could have made the network wildly popular.

Chapter 7

Key Questions That Define a Deployment

This chapter addresses some of the main issues that are, or should be, resolved in the early planning stages of the deployment after you have assessed constituents' needs. Collectively they are applicable to the municipal broadband infrastructure and to the applications that your various departments launch.

The question that has the most profound impact on the infrastructure deployment is what business model do you use. There are many options, none of which are widely proven, and the factors influencing the answer are driven as much by emotion and politics as practicality. Other questions are less volatile but still demand careful consideration. Do you build, buy or use an application service provider for your software? Do you rely on standards for hardware, software and infrastructure? Can we make the network or applications secure enough to move forward?

Some of these may seem like issues best left to IT. However, every director and most senior administrators should have input to the answers because these questions are integral to timely project completion, short- and long-tem cost containment and the ability of the technology to meet users' needs. Not to worry, you don't need to have an engineering degree to understand this chapter. But, as I mentioned before, familiarize yourself with the basics of the technology so you have a better grasp of the issues you face and the decisions you have to make.

Which business model is best?

With the municipal WiFi infrastructure, there are management issues that you must resolve, such as who owns the network, who operates it? Do we charge some citizens and businesses for access, but not all? Politics, passions, finances and evolving technologies play a role. What do you do?

First and foremost, "you have to know what's driving this interest from constituents' perspectives," states Dianah Neff. "Do you have a digital divide? Where are your schools on the issue, is wireless important to them? If you don't have a digital divide, maybe the private sector model is the way to go. Let incumbent providers decide if they want to bring in WiFi or not. If you're in a rural community, and you can't get anyone to come in at all even if you beg them, you may have to step up as a municipality to take this role. But make sure you have quality vendors working with you and you have identified what their needs are since you want to create a win/win situation for everyone."

In broadly defined categories of business models, you have the consortium model in which a group of companies come in, build the entire network and they own it and operate it. Several of these bid for the Philadelphia project. Another option is to have the local government and a business (i.e. service provider, infrastructure vendor) form a partnership in which the government and private entities share operating responsibilities. In the third category, the city owns and operates the network the same way they would a utility. A variable that comes into play is whether or not you sell access to the network. Some cities own the network, but don't charge for access.

When the Executive Committee started writing the business plan, there was little hard data, established business models or metrics to provide the empirical financial justifications that citizens and cautious politicians like to see for pricey projects. But they still made a good effort at analysis in this area in spite of so little with which to work. They called in Dr. Munir Mandviwalla, Chairman of the MIS Department at Temple University's Fox School of Business and Management to head up a research effort that ran concurrently with the focus groups. His team consisted of faculty, individuals from the wireless industry and students.

"We did a requirements analysis of the different stakeholders and their competing interests," says Dr. Mandviwalla. "We did some secondary economic and demographic analysis using census information and some proprietary survey data to see what the demand might be the Philadelphia area. We said, if you have so many people with Internet access, and you have 'such and such' presumptions about household income and 'such and such' education, let's make a guess about how many people would be interested in the service."

Working from here, the group projected how many people will be using service in one, five and 10 years. This was hypothetical since they had to make so many assumptions, but they just wanted to get some idea of what the potential was. Part of the team analyzed wireless technologies used in different cities' pilot projects and other members analyzed various business models. They also spent some time looking at the competing interests of the incumbents.

"In the end we concluded that all of those models have merit, but the potential for success is in the execution," states Dr. Mandviwalla. "You could almost flip a coin, pick one and the thing that would matter most is the execution. Management could be put together in such a way that they get the network done on time with reliable access and good customer service." There's nothing more concrete to work with because no city has actually completed a project to the point of getting everything going.

In two years cities might be able to say here is the best model, but even then it would be contextual because every city brings so many unique variables to the table. There are all kinds of variables, a lot of complexity. One thing is clear from the research group's perspective. "The government, the private sector or the community can't do this job alone, and it's painful for all three to work together."

Meanwhile, participants in the focus groups were telling the Executive Committee that on one hand, they wanted the network's operation to be outside of the political process and they didn't want to use taxpayer dollars, which eliminated the "free utility" option. On the other hand, they wanted the project to have a social conscious and help close the digital divide and bring economic development to neighborhoods. The Committee compared these requirements against different models and asked "where do we get the biggest bang for the buck?" At this point the "Wholesale Model" concept that was developed and trademarked by Civitium became the leading option.

The wholesale model takes hold

Dianah says that in this model, we created a non-profit organization to be funded by loans or bonds to pay for the buildout and focus on the social programs to be addressed through the network. We wanted to use the skill set of the private sector to design, build and deploy the network and have ISPs [a "lead" ISP and others interested in participating] pay wholesale rates to use the infrastructure to deliver the end-user service. During the contract negotiation a lower rate would be set for low income residents and businesses."

With this plan, there was no liability to the City for the build out since Wireless Philadelphia would be independently funded and operated, though obviously influenced by the City. But liability comes with the issue of granting access to assets owned by the city. This makes Philly a landlord. Some City Council members also worried about what happens if the ISP abandons the network and the city has to take over management of it. A lot of the liability discussion was quelled, though, when Earthlink in the RFP process offered to build the network on its own nickel, and the wholesaling component would stay the same.

Reading newspaper accounts in the weeks following Earthlink's offer, a fair number of people seem to be happy with the new deal. However, arguments against the project from a few opponents have shifted from "this is going to be wildly successful and provide unfair competition to the telcos" to "it's going to fail because they can't sell enough subscriptions to make a profit." Critics support this latter argument with the fact that service providers in the area haven't been able to garner more than 13% of the market for high speed Internet access, so Wireless Philadelphia's revenue estimates are too optimistic.

Executive Committee member Ed Swartz feels that "this argument about penetration rates is silly. To say that we've never received higher than a 13% penetration rate up to now is like someone saying in 1948 we haven't achieved more than 15% of households owning television sets. It's an accurate statement to make in 1948, but it doesn't tell you what will happen in 1953 and 1956." Dianah believes a municipality that owns the network and services which are provided to citizens and also has city mobile workers using the network to offset their operating costs shouldn't be adversely affected financially.

To get an idea of how the "public utility" business model might be received by communities in other cities, look to St. Cloud, FL. Mayor Sangiovanni reports that since the city's putting money back into people's pockets through free access and no tax dollars were spent because the money is coming out of the economic development fund, it wasn't a hard sell. It's easy for the average citizen to see what's in it for them. St. Cloud, by the way, is outsourcing the entire customer service operation to an ISP.

It's probably safe to say that the business model question will be decided on a city by city basis for a while longer. Bailey White, president of Pervasive Services which is providing content services for the Philly project feels that "the smaller cities might look to do just a right-of-way arrangement in which the provider really owns the network soup to nuts. The large cities and the areas where broadband service is really bad might want to play a bigger role both in setting up the network and determining the ways it is used by constituents."

However, Committee member Robert Bright poses the question, "Are we over-working this issue? We're not talking about a lot of money here, just $15 million [the proposed cost to build the network before Earthlink's proposal]. We just bid on two stadium projects costing hundreds of millions. What are the benefits of the Eagles playing in a new stadium versus the benefits of wireless? You take a large corporation, whether they're tech based or not, and somebody's going to wake up and say, we're talking about a little bit of money for a lot of coverage." How you answer this question depends on whether you approach it from the creation orientation or the problem orientation perspective, don't you think?

Two pivotal technology choices

As you move down the path to deploying wireless infrastructure and mobile applications, there will be some forks in the road. Choices that you need to make about the direction you take the deployment. There's no universal right or wrong answers. Weigh your decisions carefully because they will have a profound impact on the functionality and profitability of the finished application.

Buy it, build it, rent it?

This question applies to mobile workforce applications. You can buy "off the shelf" software that is ready to go, and possibly with features that allow you to modify some of its functions based on workers' needs. Your IT staff can use software tools to build an application from scratch. Or you can "rent" a solution for a monthly fee per use from an application service provider (ASP) that runs everything from Web servers that they own and maintain. Some ASPs throw mobile devices into the bargain as well.

The main argument for buying an application that you can use right out of the box is that it can save you the major software development costs of having your IT staff, the vendor or contract programmers write the application. The argument against is that you may have to conform your business practices to adapt to the software rather than having the application work as you do.

You have a middle ground "buy" option, which is to buy an application that is partially customizable, such as what Philly's L & I Department did. You'll likely get something closer to what you need without dramatically changing your business practices. However, if you're not clear on what you want the application to do or you don't properly manage the vendor doing the customizing, you can spend a fortune and still not get everything you want or need. For small businesses and community organizations, this is an important consideration because they may be the least likely to have the people on staff who can manage such a project.

If you don't want to buy, the main argument for building an application in-house from scratch is that, similar to a tailored suit or dress, the final product can closely fit the uniqueness of your organization and its needs. The primary opposing view is that if you don't have a sufficiently large enough or skilled enough development team in-house, or you hire an inept outside group to do the work, this will cost a lot of time and money and you'll likely endup with a poor system. A way to make the "build" option more viable is to get good mobile software development tools such as those offered by AppForge (www.appforge.com) that gives you blocks of pre-assembled program code. This accelerates development and reduces programming errors.

Some organizations (particularly smaller ones) that are leery of the cost of buying an application and don't have the staff or time to build an application will go to an application service provider (ASP). Setup costs are low, monthly fees are manageable, there's little or no system management overhead for your IT group, you can usually get out of the deal quickly if it doesn't work and sometimes hardware is part of the deal. Probably more organizations would use ASPs if they weren't worried about security issues and losing control, even though a lot of these fears are overblown.

One big advantage of using ASPs for basic applications, regardless of your organization's size, is that you're able to test the ROI assumptions behind the application, in essence creating an inexpensive pilot project. E-mail and PIM access, mobile worker dispatch, time sheet submission from mobile employees and community portals are the type of basic applications that make sense to get through an ASP.

Factors to consider

Here are a few factors to consider to help you decide which options are best. First, what is the nature of the deployment? If you're providing simple access to the Internet, e-mail or data stored in a business application, there's probably not a huge need for customization and something off the shelf will work. Organizations that require a complex application that reshape their business operations may need to build it or find software they can customize.

What's the software skill level of your development team, specifically in the area of mobile applications? How big is your team? If they don't have a lot of skill in wireless or a lot of people with the time to devote to development, building your own is likely a bad idea. Outsourcing is the way to go, and if you don't feel your in-house team has the right skill set, get a consultant to come in and manage the outsourced project. If you have enough people with the right skills, customization in-house is a viable possibility.

With hosted applications, two main considerations are: 1) can you reap the benefits you want from the application, and 2) are you comfortable with it sitting on servers other than your own, or having data pass through another company before it gets to you? It's an option worth serious consideration. If the application doesn't work out you can stop the service without losing a large up-front investment, maybe no more than a couple of thousand dollars.

What are your choices in application development tools? If the application comes with built-in development tools or you can find a good stand alone tool, consider the "build it" route. As municipal broadband deployments pick up steam, there may be a greater number of development tools to choose from and they may be more sophisticated.

While conventional wisdom is that you adapt mobile applications to how you do business rather than the other way around (the argument for the "build it" option), Greg Lush, CIO of the Linc Group, argues forcefully for rejecting that idea.

"When people talked to us about giving workers a mobile application, the assumption was that we had a good system in place to begin with," says Lush. "But the business process was not good. In fact, it really sucked. The last thing I wanted to do was automate that. I don't want to work crapier faster. We changed the way we worked by creating applications for 18 different existing business processes: managing scheduled work, dispatch, ordering parts and so on." To create the right application, management met with a representative number of field workers, described how these processes worked effectively in other places, and then asked how our people were doing things.

Next they went out and looked for a software application that rigidly adhered to most of the ideal business practices they wanted to adopt. "We like applications that are robust enough so that we can personalize within the top layer of the software without having to change the core application. Siebel has a good product, but it's 20% foundation, 80% customization. I'll spend more professional services dollars because I have to customize that 80%, plus I'll end up automating my existing process since I can 'customize the application to work the way we do.' And we've already decided that that's a waste. We're looking for 80% foundation that represents the best practices for our business."

L & I's Computer IS Manager James Weiss went the 80-20 route when he selected an application that was mostly a complete code enforcement package, and then worked with the vendor to customize the remaining features. However, his was a choice was driven more by a dearth of options. "The vendor we chose, Hansen Information Technologies, had done a couple of smaller projects with us. We reviewed the possibilities we had for building onto our legacy system. We looked at all the people who were selling apps in our government marketplace for code enforcement. We talked to comparable cities that we had on-going communication with because of our needs and interests. We found that this company was selling to almost 100% of the comparable jurisdictions with similar needs."

Tom Baumgartner, Network Administrator for the Pinellas County Sheriff's Office in Florida, was one of the pioneer users of WiFi, so he had even fewer options. "We needed to deploy two applications to track inmate movement and electronically record data to inmates' medical records. The system needed to be robust so users could obtain this information at any time in any building within the complex. At that time we found that there was nothing available 'ready made,' so we brought in a company to develop jail management software to work with mobile devices. Then we added software developed in-house such as Voice over IP (VoIP). We're planning on adding video capabilities to augment VoIP."

Should we standardize or use proprietary systems?

The question of standards applies across the board for wireless network infrastructure and application deployment. It comes down to 1) do you buy products that conform to one main wireless or computer industry open standard, 2) do you buy products that conform to a variety of industry standards or 3) do you buy products from vendors that use their own proprietary standard?

For example, WiFi, though there are different versions of the technology, is a prevalent standard for networks that move data from one point to another wirelessly. With PDAs, there are several standards for the operating system that drives these devices including Palm, Pocket PC and RIM's Blackberry, and in many organizations all three types of devices are in use. Or you have several companies that offer RFID applications are proprietary in that only radio receivers sold by a vendor will work with that vendor's RFID tags. Any upgrades or add-ons to the system have to come from the same vendor.

Standards affects costs in two ways. The more hardware produced that meet a standard, the less they cost due to economies of scale in munufacturing. A decade or so ago, an Apple Mac cost a couple of thousand dollars more than a desktop supporting the much more widely adopted Windows PC standard.

Also, when particular hardware products become the standard, there are more software applications and development tools written which support them so there's less of a need for custom software. It costs less to support hardware products that conform to one standard because IT doesn't have to learn or "hire in" different skills sets, provide duplicate sets of replacement parts, development tools, etc. Finally, if a vendor with a proprietary standard goes out of business, you won't be able to get support, upgrades or replacement parts when something goes wrong.

Adherence to standards affects how well everyone connected to the network can communicate with others and access or send data. Otherwise you may lose the ability for those across the workforce or community to communicate fully if the various products don't support the same standards. Files sent to a Palm device may not be viewable on a BlackBerry. Or infrastructure bought today that doesn't support common standards may not work well with technology you buy two years from now, causing communication breakdowns. Many cities are finding that the proprietary wireless radio networks they own, particularly in public safety, are islands of technology that can't easily be integrated with each other and with new technology.

Decisions need to be made about standards for all aspects of the wireless equation. With the municipal network, ask vendors of infrastructure products which standards do they support and what is the life expectancy of the standard? Then go talk to industry experts to have them assess vendor responses.

Dianah and her team spent quite a bit of time talking to industry experts about where they saw the technology moving. Intel was one. There were questions about what the life span of WiFi is and who will support devices that meet this standard. Because WiFi is so ubiquitous, and other mobile technologies have been slower to deploy, WiFi has a 10-to-15-year expected life span. Given that your mayor's office changes hands every few years, it's critical that you make decisions that can survive a decade or so at least to be able to withstand potential changes that come with new administrations.

Rizwan Khaliq, Global Business Leader for IBM Digital Communities, believes that industry standardization is going to be a big benefit to communities. "You want to look at how well these products are going to be 'future proof' or scaleable. On the network side, you have WiMAX and the like. On the embedded technology side, you have standards which are becoming increasingly important when you want to use the network to do things such as remotely monitor bridges, highways and the like or even business computing processes."

In terms of software, here too you want to select products that are standards neutral or support the prevalent standards when it comes to hardware operating systems. Mobile middleware, business applications and systems management applications that support the major PDA operating systems, laptops and desktops are your best bets. Your main software can keep operating year in and year out as hardware models come and go.

It also helps if your mobile applications store data in a standard format that can be shared by all or most of your other business applications. In Philadelphia's master technology plan, all of the appropriate applications are sharing standards so that city employees, citizens and businesses can all benefit from being able to access the same databases.

Mobile devices create an exception

With all of the hype surrounding wireless technologies, you don't want your organization stampeded into decisions that don't adhere to standards. Having said that, adhering to standards for mobile devices is the one area where organizations have to be flexible. One reason is that the nature of different users' jobs across the organization dictates that they have different devices. The Mayor's cabinet may be thoroughly satisfied with BlackBerrys, but inspectors who have to weave through construction sites need ruggedized devices that can survive the wear and tear. Social workers for a non-profit that contracts with the city may need full laptop capability while health inspectors do fine with smartphones.

There is also the issue of individuals' preferences. PDAs and smartphones are more personal than other computing devices because of their size and the fact that most private schedules and e-mail are accessed from these along with business information. So IT discovers that many devices find their way into the organization without their knowledge or approval. There are various ways you can handle this, from trying to mandate what types of devices you pay for to having an "anything is fine" policy. It is important that you tackle this issue sooner rather than later. People are buying, upgrading and loading devices with data all the time. That's one train definitely leaving the station without you if you're not quick.

Cities and counties find different ways to address the standards issue. Robert Smallback, Jr., Senior IS Manager for Lee County Port Authority in Florida states that "we believe in standardization within reason. The IT team watches for new innovative technology and when it arrives, we test it. If it performs to our satisfaction, it becomes the new standard. Standards allow a limited IT staff to resolve client problems quickly, so our staff feels comfortable with customer service. We only need to be experts with the standard and not experts to unknown products."

Robert's team rotates hardware through the organization following a formula where every year they buy 1/3 of their inventory and ripple the new product to the power users. They donate the low-end 3 ½ year-old products to the local school board or auction them to benefit our Police Explorers youth group. This results in a drastic reduction in hardware failures and improves the ability to adapt a new standard. Hardware is constantly changing, so they have to change to keep pace with new technology.

Baumgartner follows the approach that "you should try to keep as much as possible to standardized hardware. I feel this is very important when dealing with existing network infrastructure. This also helps in problem solving which may arise. I know where everything is and what everything does, not so much with mobile devices, but with the access points. Now, this is a great thing to do, but in the real world it does not always happen. That's why I emphasize having a good network management application that is multi-vendor friendly. As new people come on, the system has to adjust."

Security – boogie man of the under-informed, handmaiden of the obstructionist

"A lot of security consultants scare the crap out of people. And sure, if you do things wrong, security breeches will happen. A dedicated hacker can get in. But really, how often is that going to happen? Worry enough – lock your door, set the alarm. But you don't need barb wire and machine gun turrets." This statement from Bill Brook, Director, of Information Technology Chicago's Children's Memorial Hospital, sums up the security discussion pretty well.

This is a pretty vital issue because security is at the heart of a lot of resistance to wireless, both from within organizations and from those whose interests are adversely affected by municipal broadband. Organizations are hesitating in moving forward primarily because of this concern. I recently listened to a representative of a telecommunications industry association on radio deriding the value of muni WiFi, stating that no emergency first responders would rely on WiFi because of the "lack of security." The next time you hear someone give that argument, relate the following story to them.

Morrow County, where security failure isn't an option

In Morrow County, Oregon you will find the Umatilla Chemical Depot that holds about one-third of the U.S. remaining stockpile of warfare materials. This county is home to the Hanford Nuclear Reservation, and operates a nuclear power station. One of the major east-west rail lines in the western U.S. runs through this county. Morrow County hosts major natural gas and energy production as well as distribution facilities. Talk about a hot spot!

The Morrow County Emergency Management Center's (MCEMC) team of first responders relies on a 1000-square mile WiFi network to manage a myriad of monitoring and emergency response resources prepared for a potential nightmare of a catastrophe. And yet no one loses one night's sleep over security of the network. Your city should send every critic of WiFi security to spend 15 minutes in Morrow County. While you're at it, send the directors of your emergency response teams out there to see some of the interesting Homeland Security-funded initiatives they have in place.

Cameras linked to the network stream real-time, full-speed color video to monitor all of these facilities, and they can be remotely controlled to turn and zoom in on specific areas. The same type of cameras monitors the highways since in a chemical disaster the staff has about 10 minutes in which to respond. If they need to quickly evacuate residents, MCEMC relies on those cameras and the network to remotely re-direct traffic by controlling traffic lights, drop arm barriers and billboard-size digital message signs that can post new text as needed (FEMA, are you listening?).

MCEMC deployed WiFi access points mounted on buoys on the rivers and waterways to provide warnings to watercraft as well as back up to land-based access points. This same system can operate un-manned fire boats to fight hazardous materials fires on or near shores. The main PBX phone lines all have VoIP capability that also provides backups to the cell phones.

Emergency response vehicles are equipped with mobile access points so they can stay connected to the network while driving up to 100 mph. The network is HIPAA-certified safe so that patient data can be wirelessly transmitted while en route to hospitals. On top of that, the network also complies with Federal Information Protection Standard (FIPS 140-2). Passing muster for both of these intensely high levels of security is equivalent to parting the Red Sea and walking on water. All of this sort of begs the question, why is it MCEMC doesn't worry very much about security being breached, or the network failing at a critical time?

When MCEMC Director Casey Beard was an Intelligence officer for the U.S. Army, someone in the military went into hyper-security mode because it was possible, with the right equipment, to monitor and capture a person's PC keystrokes in order to figure out passwords and such. "So we had all of these lead shields brought in for the walls, PCs were put inside special metal boxes and so on. Then one day I was thinking about it and realized that, sure, you could intercept keystrokes. But you'd need a lab environment to do it, which would require a semi trailer full of high tech gear and a bunch of antennas on the outside and it would have to be parked almost in front of the building. Something like that sitting outside would be pretty obvious, I think. Historically with technology, the threats are often greater than the reality."

Casey believes that any network is as strong as its weakest point, so his team works at keeping the weak spots strong. They bought a couple of commercially available WiFi security applications to provide the HIPAA and FIPS security, and later hired a security team to come in with special equipment to try to hack the network, which they couldn't do. They constantly follow all of the standard security practices such as hounding employees to secure passwords, not link into access points without the appropriate security features enabled on their devices, etc. "There's always going to be a tradeoff between being able to use information versus protecting it. If you want total security, then don't use technology at all."

To address the issue of what happens when natural disaster strikes, Casey's crew relies on not one, but two massive fiber networks that come in all the way from Portland. The WiFi mesh integrates with the fiber networks and also has the access points densely deployed so as to provide overlapping coverage in many areas. There are trickle charge batteries that back up access points for eight-totwelve hours and in some case solar panels recharge the batteries, plus uninter-rupted power supplies and standard generators provide yet more backup. The access points are placed to minimize vandalism. Don't forget about those mobile access points providing yet another level of redundancy. If all else fails, Casey tells me they keep an adequate supply of white boards and grease pens on hand.

In that same radio interview the telco industry rep also was fond of saying that WiFi wasn't more reliable than any other communication option because when Hur-ricane Katrina hit, all of the communication structure collapsed. This is true. But the day following it, the mayor managed to get in touch with the world by going down to an office supply store and borrowing some WiFi equipment. WiFi might have col-lapsed along with everything else, but it was also one of the first things to come back online. Roving teams of volunteers in various parts of the devastated areas brought in WiFi equipment and VoIP gear to connect displaced relatives with their loved ones. Even when cell phone calls weren't getting through, text messages were moving in and out.

Knocking down the Boogie Man

Believing that boogie men and obstructionists should be cast from our midsts, here are some knowledgeable words on the subject of security and some tips on how you can make good security decisions and move on. I contacted Jeb Linton, Chief Architect for EarthLink Municipal Networks, with a couple of theories I wanted to run past him.

At the fringe of the network where access point meets computing device, isn't security in the hands of the user as much as it is when that person's using a wireline connection?

Yes, though it's possible and advisable to give the user more tools for security than are commonly available on wireline connections. Most network op-erators should make an unencrypted hotspot-like service available for casual users, but they should certainly make a high-security WPA mode [Wi-Fi Protected Access - a security protocol for wireless 802.11 networks] available as well. Failing that, even providing a VPN service would be a viable option. Shared-key WEP, on the other hand, is completely unsuitable for a public access network. In any case, there should certainly be an attempt to make the user aware of his/her risks and options using a captive portal splash page.

Conversely, if a city's servers (or any Web or intranet servers) are adequately secured, are they any more vulnerable to hacking from someone with WiFi access as landline access? I realize that finding a hacker may be harder if they use an access point, but that's another issue.

Aside from the issue of finding the hacker, as you mention, there's no functional difference between the two scenarios as regards vulnerability to hacking.

What are the main issues being distorted in the attack on muni WiFi?

The issue of interference, while worthy of careful consideration, has been blown out of proportion in some places. Intentional interference with an unlicensed-band wireless service is quite simply illegal and the FCC won t put up with it, especially if it's done on such a scale as to actually put a dent in someone's business. Unintentional interference will certainly be common but a well-engineered system can deal with it quite adequately to the point where it s unlikely to be noticed by many users.

What are the top five things a city can do to make municipal WiFi secure?

- Give the user the option to use a strongly encrypted WPA connection.

- Make the user aware of his or her level of security, in plain language.

- For users that prefer to attach to the network unencrypted, allow and encourage them to use VPN software.

- Over-engineer the WiFi network for *both* coverage and capacity in order to deal with interference, and be sure to budget for upgrades over time.

- Don't allow free access areas to compromise security. Log user sessions and take precautionary measures such as port-specific rate limiting, session time-outs, and unauthenticated-SMTP port blocking to avoid attracting spammers and DDoS attackers.

What are the top five things a citizen (or city employee accessing the city's network) can do to ensure data security?

- Use a secure WPA access mode if one is available.

- Use a VPN product if you aren't on an encrypted WiFi connection.

- Don't rely on shared-key security methods like the original WEP systems. Remember Ben Franklin's comment: Three may keep a secret - if two are dead.

- If neither WPA nor VPN options are available, don't send or receive sensitive data unless you have application-level security. This means checking the little padlock icon on the web browser, for example, to make sure the web page is using SSL/TLS security. The easiest mistake to make here is not using an encrypted connection to your e-mail server. Most decent services have a high-security mail option.

- Practice good password management habits. At minimum, choose passwords that aren't obvious (following the standard guidelines), change them regularly and never share them with others.

What is Earthlink doing to ensure security of the network?

Earthlink will be providing CPE devices to our customers which will automatically attach to the network using a strongly encrypted WPA mode. We will also provide WPA supplicant software for those users that wish to attach using other WiFi client devices, and will provide users of the unencrypted access option with information about how they can surf safely. We will actively and continuously monitor the network for interference and attacks, and will continue to update our network and supporting systems as security methods evolve over time.

Bottom line? If you adequately secure the city's computing resources and businesses adequately secure theirs, then your vulnerability to WiFi surfers isn't much greater than it is from wireline surfers. If you don't secure your servers and access points or monitor effectively for rogue access points, then bad things will happen. The average citizen is going to going to be a weak point in their own security picture as long as they don't enact safety precautions on their computers and mobile devices, which you can encourage but probably shouldn't legislate. The potential for laws to do more harm than good in this area is pretty high.

Address security and get on with the business at hand

Morrow County isn't the only area where people have made sensible security decisions. At L & I, "wireless communication is done to fixed IP addresses from the devices," says Weiss, "so we use that restriction to make sure that whatever people try to do to in terms of getting into our system through channels that are open to inspectors is ruled out. To date we've had no recorded incidents of anyone trying to get in or any questionable corruption of data that might be attributed to something that would happen this way. Internally, a big investment has been managed by MOIS for firewall and security for outside transactions. That part of the project has not been problematic, but we're always vigilant."

There's no permanent storage of data on devices. It's there temporarily while inspectors goes through a site or if they have collect information and don't have a connection at that time. Once the communication has been confirmed by the device shaking hands with the intermediatry server, then that data will be erased on the device and the replicated data moved to from the server to permanent storage on a production server.

L & I supervisors know what inspections their people are doing even if they don't have a signal, and if something does happen to a device the IT staff is able to reconstruct what was on it. When someone calls and says they can't find their data, IT can restore the user's system configuration and data, and then use a wireless communication session to have that person back in operation in about an hour.

Robert Smallback's team at Lee County Port Authority identifies what security areas they want to address and let their service provider implement the system to those specs. A test system is configured to those specifics, installed and then tested by a third party who validates everything, identifies anomalies and makes recommendations.

The staff then makes additions, changes, deletes and alterations to the systems. They are trained by the provider, and act as the first line of support. Security analysts from Cisco, SBC and Siemens perform hacking drills once or twice a year, and the office follows their suggested fixes to our vulnerabilities. Robert varies the firms to keep the security tests pure of preconceived concepts.

"Recently our wireless application was tested by the local TV investigative reporter who hired two hackers to challenge our site. The report was excellent, showing that our wireless network was very robust and easy to use. They could get into wireless devices used by passengers but the hackers rarely could see the Port Authority's network. And when they did, they couldn't penetrate it. If we had developed the system in-house, and configured it for security, then tested it ourselves, we would not have gained the value of the service providers' expertise. You can do it once the best that you can, or you can turn the job over to people who do it for their daily bread. I'd go with the experts."

Baumgartner advises that you plan constant security testing procedures to follow when you add new applications so that you're always looking for vulnerabilities. "Commit to documentation to make sure everyone knows what to do. To support our mobile devices, we're adding authentication servers. We have to adhere to Florida Department of Law Enforcement hardware requirements for any devices that connect to them. Keep vendors close at hand who can help with problems if they come up, and regularly check out what new security features are available.

At the San Diego Fire Department's Medical Services units, the devices are always under control of the crews in the vehicles, so they rarely leave the possession of one of the personnel. What's more, there are never more than one or two reports on the device at any time even if it were to get lost. There was actually more data security risk with paper forms because our people were traveling with completed forms on the dashboard. When crews return to the station, the data is immediately uploaded and only administrators and the person providing care to the patient has access to data.

There's not a good way to put a physical leash on devices because of how crews work. But an Intellisync Data Sync software feature lets IT remotely disable the ability of a stolen device to sync with the server. They thought about password protecting the devices, but felt it was too much trouble relative to the risk. As the pilot continued, the project team didn't find any significant risks of patient data being compromised given the practices they have in place. If the devices were more expensive the department might worry more about theft and put more restrictions on their use.

Dr. Joyce Copeland, Medical Director of Community and Family Medicine Physician Assistant Education at Duke University Medical Center adds another security option. The database is designed with codes so you can't figure out who the patients are since the data is so vague that it's difficult to trace back to specific patients. In some respects, the types of details that the hospital needs are so general that residents don't ask for sensitive data to begin with. They want to know what types of procedures residents performed and how many of them, not the name of the patients they performed the procedures on.

Dr. Copeland says "we joke around here sometimes that, if there's a way to screw up the data the students will find it, so that further ensures no sensitive data will escape. But on the serious side, people are skittish about security so we include password locking on all of the laptops, PDAs and servers. We encourage backing up and students are given backup cards. If they do like we ask, we can find and recover any lost data for them."

Security is indeed a serious matter. But as Brook concludes, "Wireless networks are inherently less secure if you don't do it right. People can drive by and pick up your network. So our wireless networks are on VLAN. We're following all best practices. If you have confidence that you can secure your network, then you will be comfortable. But there are CIOs who are not technical, or CFOs who've seen these Tom Cruise Mission Impossible movies, so they fear the worst." Don't let the Boogie Man cloud your decisions.

Tips on department managers' role in security

In many organizations, security is one of those "geek" issues that executives and end users alike want no part of, but will complain bitterly when IT imposes what are perceived to be unreasonable guidelines. However, with mobile devices playing the role of pocket-size anywhere, anytime gateways into the organization's data resources, security is everyone's responsibility.

IT people, of course, should be aggressive in their data security tactics since they have to deal with the mechanics of implementing security technology. There are two primary concerns to address: security of data on the device and security of the back-office LAN. The most common approach is to password protect the device's data and password restrict their access back to the network, which is sufficient for many organizations. For extra duty protection, you can buy encryption software, though this will add to your TCO if you have to pay a per-user license fee.

Ellen Daly, Principal Analyst for Forrester Research, observes that "people also are starting to put software in place that scans any mobile device that connects to network to see if the device is compliant with security and software licensing policies before allowing data to be accessed or downloaded. If the security software sees something out of line, it will shut out the device."

Developing strict security policies is easier when the devices are restricted to business use, such as inspections or asset management, assuming users don't have to learn complex technical procedures. But in a deployment where the devices such as PDAs and smartphones contain both personal and business data, they become so personal to the user that you risk a backlash if policies are viewed as heavy handed. Forrester recommends that IT models guidelines for PDAs and smartphones that are similar to those for laptop security, including rules and procedures for backup, access restrictions for lost devices and how data will be deleted when employees leave the organization.

The business manager's role is to create a "culture of security" among employees. In this culture, it's ingrained into the mobile workforce that any laptops, PDAs or mobile phones that store and access company data are government assets. Senior management makes it clear that IT's security guidelines, passwords, access procedures and the like are to be followed as if users' careers depend on it. Reinforce these lessons with penalties when personal security lapses result in data loss or a security breech. Few people want to leave their year-end bonus payment or promotion sitting on a subway car.

Be sure your culture of security encompasses the use of WiFi. You should have a set of on-the-road security guidelines, including the proper use of devices when accessing hotspots or even access points within the city's network. In-office access points are potentially windows into the soul of your network, so directors and managers must be involved in the decisions as to where they are deployed and how they are accessed. Caution must be second nature.

Ultimately, at the end user level is where most security policies succeed or fail. Employees and citizens either will follow guidelines or they won't. You can't chain devices to users' wrists, or be standing over their shoulder to make sure some nefarious type isn't watching over the other shoulder. Once you make security as airtight, user friendly and comprehensive as you can, educate your workers and your constituents. Then (hopefully) good sense will take over.

One last piece of advice. Whatever security procedures and features you put in place for emergency situations, make sure that your administrators and managers practice using them once or twice a year. It doesn't do you a lot of good to have all of these emergency response actions planned, but no one remembers the procedures by the time a disaster strikes. Readiness is all.

To sum it up

Moving forward to deployment involves resolving a lot of questions and concerns by city workers, citizens, vendors and others involved in the process. There's definitely no such thing as a dumb question, especially when the right answer can save you tens of thousands of dollars, schedule interruptions and lots of aggravations. This chapter tackled only a few, albeit very important, questions. The remaining chapters help you address others, starting with selecting the right technology for the job at hand.

Chapter 8

Aligning Technology Initiatives with End User Needs

Making sure the network infrastructure you put in place aligns with the needs of city workers and various constituent groups is a top priority. However, you also must be prepared to influence how technology initiatives within the departments and communities align with needs during the initial months of the network's deployment, even though you may not have direct responsibilities for those projects. Your success is very much dependent on theirs.

This chapter gives you a general game plan for addressing this rather complex segment of the implementation process. A number of people who played different roles in Philadelphia's wireless efforts share their insights and offer good advice to help make things a little more manageable.

Understanding your options

Let's start by looking at the main components of a typical network infrastructure and then the main product categories for mobile workforce applications. Mine are thumbnail descriptions but there are plenty of tech experts who can fill in the details.

Components of the broadband network

The basic municipal broadband deployment has three main components: the edge (also referred to as the fringe), the middle tier and the distribution tier.

The edge of the network is where the radios are placed around the city or county, typically on light and telephone poles. There may be thousands of them in a large city communicating to mobile devices equipped with WiFi cards or embedded WiFi chips, sensors placed in physical assets and any WiFi-enabled products such as cameras or computer printers. Unlike the access points for your home or office, these are more powerful so they can communicate with devices further away. Tropos Networks, BelAir Networks and recently Cisco are some of the main names in this part of the business. These vendors participated in some of the Philly pilots.

In the middle tier is a collection of gateways where all of the data gets delivered to and picked up from the access points. These may be deployed in a ratio between five radios-to-one gateway and three radios-to-one depending on how intense you want or need the coverage to be in a particular area. The fewer radios per gateway, assuming a highnumber of gateways, the more concentrated is the coverage in an area. Motorola with their Canopy systems and Alvarian are two of the major players in this part of the business, but you can re-program the on-board software of the Tropos radios, add a connecting cable and these then can serve as gateways. The combined collection of edge and middle tier technologies is often referred to as the mesh.

The distribution tier is the communication bridge between all of the gateways and the point of presence (POP) which in turn transmits the city's data to and from the Internet. For this tier of the network, vendors can create a sonic ring of redundant microwave links between six or seven communication towers. Each tower has a massive high speed link to the POP, and data flows back and forth between the towers and the gateways.

For small towns and rural communities such as those in the west and mid-west, your network infrastructure may be a little abbreviated. The density of homes is low, plus you don't have street lights on every block from which to hang access points. Here you need to look at solutions that provide access over large geographic area, which is where gateways make the most economic sense if you expand their reach directly to the buildings and users on the edge. However, the customer premise equipment (CPE) has to be more powerful and subsequently more expensive. A bridge to boost the RF signal from a home to a an access point in an urban area may cost $60 or $70. For a home or farmhouse in a rural area, they'll need a CPE that costs $500 a piece when bought ingroups of 50-units.

Mobile workforce technology

Technology for mobile workforce deployments can be divided into five categories. There are end user computing devices, including PDAs, smartphones (combination cell phone and PDA), tablet PCs, laptops and desktops with wireless cards or built-in WiFi access capability. In many areas it may be necessary for employees accessing the broadband network from a computer within a building to have a CPE. The connection from access points may not be strong enough to reach inside and pull data back out because of construction materials in the structures or size of the building.

Cell phones technically are mobile devices, but are not considered when discussing mobile data unless someone is planning an application in which only text messages are sent back and forth between mobile workers and the office. The newer cell phones can handle this. But most companies have decided that if mobile workers need any serious amount of data capability and voice together, it's more economical and efficient to deploy smartphones such as the palmOne Treo or RIM BlackBerrys.

The second category is comprised of the tiny wireless devices such as sensors and radio frequency identification (RFID) tags which are embedded in physical assets for wireless tracking and management.

Dedicated mobile software applications that enable operations such as data collection, remote data access, forms processing and asset management comprise the third category of end-user wireless. One portion of these applications runs on your servers and there is a client portion that runs on mobile devices. There may be a portion of code that's transferred to embedded devices, or vendors that sell both software and the required devices will pre-program them with the appropriate software.

The fourth category, frequently called middleware, is software that enables mobile workers to access e-mail and other back office software applications you currently use to run city business operations. Middleware modifies certain functionalities of your software so it can adapt to the small screen, operating systems and other features of a variety of mobile devices. Many back office software vendors are modifying their applications with middleware features to support mobile workers.

Finally there are system management applications. These enable IT and business managers to control security for the office network and devices, facilitate device and software deployment, control upgrades, track usage and other related tasks.

Constituent technology

Most of your constituent groups will use municipal WiFi as an access service for their computing devices and applications. Those cities that actively pursue social and economic development objectives, however, have to address the issue that the constituencies they're attempting to bring online often don't have the required technology resources to participate. This is why, for example, Wireless Philadelphia is working to get 10,000 computers donated to low income communities.

Judy Miller is president of Ninth Wave Media and a key participant in two of the pilot projects developing community Internet portals. She is passionate about the value that wireless can bring to efforts to transform underserved communities since seeing the impact of the community portals she developed. Judy outlines four main components that are the foundation for social or economic programs you may pursue.

First, you have to get WiFi-enabled equipment in as many homes as possible. Unless people can access the network in their homes, many will never take advantage of it. The second thing is having training classes so that people can learn how to use the key Internet resources the network makes available. At the pilot launch party in Norris Square, 20 laptops donated by HP were raffled off. But to be eligible there was an application process and one of the requirements was that the winners had to attend computer classes.

Third is relevant content delivered through a community portal or similar Internet-based resource. A portal condenses content to that which is most relevant for a particular group. Individuals who've had little or no exposure to the Web do best if their initial experience is focused on resources they can use immediately. The fourth element is a creative portal marketing effort. Unless people know about it and what it can do, you won't have the buy in from the community that you need to achieve the goals of this project.

These four components are equally effective for bringing disadvantaged businesses online. The training should be expanded to include lessons on how to market and do business on the Internet.

Aligning technology initiatives with mobile workforce needs

By and large, if at the beginning of a project the business side of the house doesn't drive the process of aligning technology initiatives with business objectives, your chances of deploying a mobile implementation that gets good results are 50-50 at best. You department directors own the business processes that make your government work, and they manage the employees who do the work. No one knows the intimacies of operations as well as they do.

In Philadelphia's case, the CIO is a position on the Mayor's cabinet. The Mayor's Office of Information Services (MOIS) is structured so that it provides centralized IT management to ensure that all departments' technology initiatives fit in with the City's master plan for integrating applications and centralizing databases. But each department does its own needs assessments, drives its technology initiatives and many have dedicated IT resources to deal with the day-to-day implementation issues.

If you are pursuing municipal broadband as a way to improve government operating efficiency, use a similar centralized IT group overseeing the deployment of mobile workforce and other wireless initiatives across departments. Create a steering committee comprised of managers and IT people from the respective departments that are deploying applications. Put members of the network team on the committe to provide guidance that ensures the applications operate nicely with the network as it comes online. Have this group meet regularly.

As people define their departmental needs in this group environment, other departments may see where there are opportunities to integrate functionality, tap into existing software resources and so forth. The more collaboration on the front end, the more successful you'll be with deployment. In the day-to-day execution of a deployment project, IT has to take a lot of the responsibility for keeping everyone and everything aligned, while department managers channel feedback to IT from users on the front lines.

Ten business-side tips for getting the best from your IT group

Senior officials, department heads and line administrators need to understand some fundamentals about your role in this wireless implementation dance. Here are 10 to move you in the right direction.

Effective alignment begins with your state of mind

1. <u>IT is neither your enemy nor your beast of burden</u>. As long as everyone realizes that they're all working ultimately for the organization's success (and acts like it), then you will have effective end results. IT people have legitimate responsibility for the operation and security of the technology systems that help your organization operate effectively, efficiently and profitably. They also likely know more about what works and doesn't work from a technical perspective. Find common ground to make the technology and IT resources work for you, not common cause to gripe, moan and complain.

On the flip side, you personally must be willing to build bridges, as well as treat the people across the table as equal partners. These aren't people who sling code and lay cable at your beck and call. IT folks are very creative at using technology and solving problems. Respect them for their talent and treat them the same as any director or manager sitting across from you. And whatever you do, be reasonable with your deadlines. Neither Rome nor any mobile application worth paying for was built in a day.

2. <u>You don't need to know programming, but you must understand some IT processes</u>. Spend time informally with an IT person learning about their basic issues regarding wireless technology, programming, vendor options, etc. You'll better comprehend how an implementation will achieve your business objectives. Despite the marketing hype, wireless and mobile have limitations typical of all technologies. If you have a good grasp of their limitations, you'll do a better job of uncovering the potential benefits specific to your situation because you won't expect the technology to do more than it can . Also, knowing how IT people do that voodoo that they do so well will help you better direct and evaluate their efforts.

Lean towards simplicity and prioritization

3. <u>Reduce business needs to terms your mother will understand</u>. Be simple and direct about what you want the technology to do. Few people will have trouble understanding the goal if you tell them "we need a mobile app that lets our inspectors complete and submit citation paperwork, as well as print violations at building sites." You don't have to solve every business problem in one implementation. In fact, it's much better if you try to solve one or two, bask in the ROI glory of that success and quickly move on to the next problem – as long as you have a well defined roadmap for the entire technology implementation process.

4. <u>Demand functionality a sixth grader can use</u>. Regardless of the application's complexity, the user interface must be simple. Be dogmatic about this. Nothing kills ROI faster than workers refusing to use a difficult application. Commands, screen prompts and data request should use terminology workers can understand. Most people don't know arcane technical terms, particularly people who haven't been regularly exposed to technology.

5. <u>Don't ask for it all, all at once</u>. Few things cause delays, cost overruns and disappointments like the "everything-but-the-kitchen-sync" approach to application design. Except maybe its ugly stepchild, Feature Creep – that continual adding of "gee-wouldn't-this-be-nice" features. Lay down the law, particularly to the top brass - no new features once the project starts! That's why there's version 2.0

Be clear on your business objectives

6. <u>The application needs to increase revenues, save money or help you run a better organization</u>. If you can't align a project objectives within at least one of these categories, why are you wasting IT's time? Conversely, whenever IT recommends a project, be sure they justify it along these lines. Not only will IT develop a better solution, you make it easier for everyone to measure success. Going back to item #3, printing a violation on-site is a clear definition of what you want an application to do. But if printing the violation isn't going to save time because you still have to mail it through the regular U.S. Post Office to some location, the feature may not be worth its extra cost. Or perhaps it's something that you put off until the next enhancement of the application.

7. <u>Accelerate or eliminate processes</u>. You're not wireless-enabling isolated tasks. You are smoothing out wrinkles in your business operating procedures, or eliminating processes that prevent people from getting real work done faster and more cost effectively. While it's true that saving paper and shaving time off each constituent or client visit are key measures of a wireless application's success, be clear with IT about which overarching business processes are being addressed.

For example, a business objective may be to save money by switching social workers from paper forms to digitized forms on PDAs so each client visit is faster. But the ability for PDAs to wirelessly access reams of data about potential resources for a client during the visit also enables the city to speed up the process of moving people from city assistance to self-supporting jobs. Or maybe the real business need is not for wireless access to e-mail, the oft advertised value of wireless, but speeding up reporting from the field so the city can collect federal funds faster. When IT can see the big picture, they'll develop the most appropriate application that may use a direct connection to the WiFi network rather than e-mail.

8. Improve lines of communication to various end-user audiences. Whether or not you produce a profitable wireless implementation depends on getting the majority of your mobile employees to use it. When you meet with IT to plan your implementation, clearly define a process for IT to gather feedback from end users about how they do their jobs, what their operational needs are and as the application is tested and deployed, what works and what doesn't.

Also create a mechanism for regular communication among the key stakeholders. However, open and frequent communication does not open the door for department managers to micromanage IT. As one IT executive warns, if your senior director stands up during the meeting and starts doing database design on the whiteboard, then you're in trouble.

Be clearer on quantifying or qualifying results

9. Ideally, every objective has measurable benchmarks. While it's best to be able to quantify the dollar impact of your deployment, sometimes it's enough to quantify the number of tasks reduced, street repair jobs increased, etc. The key is to have city departments and IT agree on what you're quantifying.

10. Even "warm & fuzzy" has a measure for success. Wireless may not be about the money, but rather constituent or employee satisfaction with service improvement. That's fine as long as everyone's agreed on what the objectives are and how they will be measured. Don't be afraid to set up some sort of metrics to gage these intangible successes. How many kids did you make smile? Can we double referrals to our housing development services? Is it true that positive employee feedback forms have tripled?

It takes two to tango – IT's role in the relationship

Of course, IT has a role in making sure their relationship with the business side goes smoothly. There are more elements of a wireless implementation than with typical technology applications, so IT has to work closely with managers to make sure all these moving parts are well coordinated so they come together for maximum effect. IT also has to aggressively manage expectations about what the technology can and can't do so there isn't much wailing and gnashing of teeth.

If you look to some of the commercial entities that have deployed sizeable wireless applications, there are valuable lessons to learn. Joyce Lewis, the Sr. IT Director for international banking company HSBC, followed on a rather defined process for keeping executives in sync with a major wireless project.

She used a requirements investigation process in which her team initially sat down with the business "customer" and stepped through in detail what they believed they wanted. In this stage IT just listened and gathered information, then went away to analyze technology options that would meet the needs identified. Next they did a market comparison of solutions from different vendors and evaluated different versions of one vendor's product.

"Based on these activities we created high-level project plans so our team could speak in an organized way to the main business stakeholder and the person who controlled the purse strings for that department about what the project might entail. On the tech side we would have a detail-level technician and project manager. If the business customer decided they wanted to move on the project, it would get funding. Some business unit managers had sign off authority, while other projects needed to go up to Finance and Purchasing. We would also determine if people on either the business or the IT side could implement the project alone or needed outside consultants."

If you're going to do something similar, be sure to include a clear picture of the good, the bad and the potentially ugly aspects of the initiative. Gregory Morrison, VP & CIO of media heavyweight Cox Enterprises, Inc. has his IT team educate executives on the possibilities and liabilities of the technology. He finds that when senior managers develop an intuitive understanding of what technology can and cannot do their requests tend to be more realistic. To this end, Mr. Cox's team creates a steering committee of business unit leaders and IT management.

For Philadelphia, it was difficult being the first large city to move so aggressively with their plans because they didn't have a lot of anecdotal evidence to help managers see the full range of possibilities. Now, the various departments can look to a few small towns and mid-sized cities plus a growing number of departmental efforts in larger cities to see deployments that spark ideas they can replicate in Philly while dodging potential pitfalls.

The average IT person may not want to become a master of public policy or public administration. But nevertheless, they should spend some time learning about the basics of how mobile personnel in the various departments' do their thing. Take a ride out in the field and see how jobs are done. It will improve their ability to get more relevant information and a needed reality check about operational processes.

Getting inside of each other's head

Wireless deployments on a citywide scale can take up to a year or more to complete, so have IT put processes in place so they see what the business side sees as things change. Depending on how aggressive business plans are, or how quickly needs might change as the city network comes online, determined how frequent IT and department units should meet.

Keep IT informed about what changes, including political ones, are affecting the city's operations. The departments need to know about new tech developments such as upgrades for a back office software program and new mobile devices that can improve their use of field data.

Understand the personal motivations and agendas of all of the people involved. Sometimes departments foolishly want to be too aggressive using a particular technology for no other reason than to have bragging rights. Or you have to watch out for people who'll reject the most tried and true technology option because there's a new fad that's getting air time. Wireless encourages this since so much is new every month. IT often hopes that wisdom and logic prevail, but sometimes you have to make a career-effecting decision on how hard you want to press an issue.

Business people need to understand that some IT interests are driven by technological purists. For example, look at when people began implementing UNIX years ago. It was a superb communication platform. It did exactly what it was supposed to. But it was not user friendly, and it wasn't easy to find skilled people who could implement and support it. While the tech person wants it because it's good technology, a business manager needs to intervene and show how and why it's not good for their people who have to use it."

To be or not to be. That is the alignment question

Acting Commissioner Robert Solvibile's and Computer IS Manager James Weiss worked pretty closely together on L & I's wireless deployment and they implemented pretty extensive efforts to align technology with needs. Each unit of the department identified 24 processes that they wanted to automate with the software. "First we created the 'As Is' doc to see what we were doing," states Solvibile. "We brought in all of the clerical staff and anybody else involved in the process to tell us how it worked."

While the manager and the clerical people were still in the room, they created the "To Be" document. This detailed everything the units wanted including all of the bells and whistles. Then the Deputy Commissioners and the supervisors were added to this group to get their agreement that this was what they wanted as well. Then the Commissioner and his Deputy would sign off on the final list and Hansen's (the vendor) staff began preliminary work on the software.

Solvibile observes that, "like anything else, once people see what they want implemented begin to take shape, they start asking 'what if we did this?' We'd make suggestions for improvements based on feedback while our staff went through a system of testing the 'To Be' features. They ran various examples of work projects and entered certain problems into the software to see how it reacted. And then we would change things and try a different problem."

During this process, but before moving to final programming, the department's staff and the programming staff each created a written document of understanding as to what the respective groups expected the final outcome to be. Then there was a clarity conference to make sure everyone was speaking the same language.

These procedures may seem to be excessively compulsive, but most business executives and IT managers who've been through a less than financially rewarding deployment will tell you they wished they had been as thorough. In fact, Cole Reinwand, Earthlink's VP of Municipal Product Strategy and Marketing, states that "a lot of cities that we're working with, before they even issue RFPs, have gone through a pretty extensive process of needs assessment. Some have polled as many as 100 personnel from different departments to determine the needs across the city." Nothing's guaranteed except death and taxes, but you can be reasonably assured that this degree of attention to aligning the technology to your workers' needs will save you lots of money and grief while leading to a more productive work force in the long term.

Aligning technology initiatives with community needs

One thing that will determine a city or county's credibility early in the deployment process is the quality of the network's service. As you make decisions, Paul Butcher of Intel encourages you to keep in mind the reasons you've decided to have the wireless network. "There are probably thousands of ways you can use the technology, so you don't try to do everything. What are the top choices for your constituents? Then go through your inventory of assets and determine what's there that will help you support constituents' efforts – vertical assets, fiber cable in the ground, rooftop assets. Do some spectral analysis and find out if there's any noise out there that can interfere with the technology you or constituents are thinking about using. Then pull all this information together and make final decisions about which technology makes the most sense."

Dianah adds that "because WiFi operates in an unlicensed portion of spectrum that multiple entities can use, you have to know who else or what else is broadcasting in the space where you want to put your network. Within that 2.4 spectrum there are three different channels, so you need to do channel analysis to determine how you work around other people who are in that space. You have meet with whoever else was there before you to come up with a way to share. Always be vigilant and work with the FCC to see if they make more spectrum available for municipalities."

Also important is the quality of service and support you give to the community leaders and others who become your champions facilitating deployment. Wireless Philadelphia has found this to be a significant management challenge as they address the diverse applications and programs that people need. Remember, by publicly embracing this new technology and your city's plan, these leaders are putting some of their reputation on the line for a program that may take many months before the benefits are fully realized.

While you are lining up community champions, be sure they are aware of the technology's limitations. For example, WiFi signals only travel but so far. They will penetrate some surfaces but not other and heavy foliage affects signals, so business offices or residences may need bridging devices. I strongly advise you to prepare written materials that outline the basic technology requirements for end users to participate on the network. Remember the rules for communicating needs to IT: use terms your mother can understand.

Patricia DeCarlo of the Norris Square Civic Association (NSCA) says "Don't tell me I need a booster. Write it down. What's it called, how big does it have to be, who sells it, how much does it costs? I have a very high opinion of myself and I assume that my stupid questions are a whole bunch of other people's stupid questions. When I go to Radio Shack, or Staples, I would not have a clue what to say to those people, what to ask for. I need someone to do a one-page flyer that tells people if you have a computer and you want to access Wireless Philadelphia, this is what you need to buy to make it work."

Taking it to the streets

To determine what technologies can best address constituents' needs, Karen Archer Perry of Karacomm believes "you ask them the question 'what is it that you want to get done?' and then work together with them to find the right answer. I described the value of broadband wireless with everyone I met, and once people understand what it was, they came up with ideas and talked to their friends who came up with more ideas. Finding a win/win opportunity for virtually anyone is easy to figure out."

When I do research and interview someone who is new to a particular technology I want to educate them about, I typically start with the question, how are you doing such and such now? As they respond I might probe a little to understand why they do something a particular way, or what would their ideal process be to do that. As they respond, you can usually find points where the technology can deliver that process, or something close to it.

An example of creating winning opportunities comes from Patricia as she explained a serious communication challenge her group has. "Getting information out in this neighborhood is a very time consuming and expensive proposition because the newspapers don't cover what's going on here, not even the local Latino newspaper. So whenever we need to be in touch with the neighborhood, which is over 10,000 people, you have to create the flyers, and distribute them. To be sure the message gets through you go out, knock on doors and explain things to people. You have to go to the corner stores and bodegas to talk to people so they'll talk to the neighbors who come into the stores later. All of that and you still don't know how many people know about the event."

Patricia and others in her neighborhood didn't understand at first what a wireless portal is. But after the representative from the City explained it was a technology tool that is specific to the neighborhood that gets the word out faster and easier than the process Patricia described, its value hit home. For them, because it was neighborhood based, portals made a lot of sense. She then was able to work with the vendor (Ninth Wave Media) to figure out different applications to create that the neighborhood can use.

Community organizations such as NSCA need to sit back and see not only what the neighborhood wants, but also how the organization itself can use wireless to facilitate the services that they provide. How can their organization grow and become more efficient using the network? There's a lot of thinking and a lot of asking. How are they going to pay for it all of this technology? How do these portals get maintained?

As she learned more about the possibilities of wireless, Patricia started thinking of ideas to help the association. "We have the child care center and the after school program, but people need to come over with paperwork to enroll their kids and hope that the person they need to see is there. I'm wondering, would it make sense for people to enroll their children in these programs, or make appointment with the housing counseling people over the Internet, and not have to depend on coming here to complete paperwork? This is something that would work for us and for folks in the neighborhood."

The big juggling act

The obvious challenge to cities and counties is that you need these community and business organizations to carry much of the load putting the right technology in place locally, but a lot of them have difficultly doing this on their own. They may not have the technical skills on staff, or the money in the budget to make it happen. While you have meetings after deciding to pursue muni WiFi, give thought to these challenges. Most likely the solutions lie in pulling together the right team of vendors and service providers who can work with your point people in the communities.

As your people work with the various constituencies, be sure they encourage groups to prioritize the many ideas that come up, giving preference to those that can be implemented easily and with easy-to-achieve benefits. The larger the municipality, the more diverse is the range of products and services running off the network, so simple initiatives minimize the strain on your employees, partners and the infrastructure while it is being built out, tested and refined.

Lutheran Children and Family Service, similar to many businesses, has several hundred mobile workers who travel primarily within the city. In the beginning they didn't know quite what to expect from muni WiFi or how it would work. Now there are new ideas coming up every day about how to use this technology. Brian Loebig observes that "since it's wireless, you can do things on the street to implement programs that aren't dependent on people being plugged into a wall outlet. You can carry laptops and palm pilots out to the clients. It creates new ways of thinking about technology and the services we provide. I just bought a wireless hub so interns in the office can do things such as research online during the day without using our main network."

Executive Committee member Ed Schwartz believes that the momentum of having the whole city deploying wireless spawns subsidiary projects that in and of themselves are beneficial, such as those in Norris Square and Olney. "These types of projects open the doors to other creative ways to use the technology, and frankly, other sources of funding because those projects were not funded by local government. They were supported by national corporations that wouldn't have touched Philadelphia before any of this. I also think that a project such as Wireless Philadelphia which is cast as an economic development initiative can be a lure for federal and state contracts."

Exploring technology alignment options

Every city and county will have to uncover what type of applications and content makes sense to benefit their various constituents, and subsequently what type of technology will need to be in place. This section addresses four areas of community interest that likely will drive projects and gives you some food for thought.

Improving health and public safety

At both the city and county levels, consider seriously how broadband wireless integrates the city public safety organizations such as police and fire departments with their county counterpart agencies, as well as hospitals and other medical facilities. As we saw with 9/11 and four years later on the Gulf Coast, how well these entities coordinate communication amongst themselves impacts the quality of service that health care providers deliver.

On their own, many health care facilities are deploying wireless in one form or another to improve how they deliver and manage patient care. Some digital divide initiatives are designed to improve citizens' access to important health care information, and improve medical professionals' ability to communicate with citizens.

The next step is to leverage the municipal broadband network so health care providers can be a better resource to public safety personnel in the field, especially during large-scale emergencies. Ambulance services such as American Medical Response and fire and rescue personnel such as those in San Diego are integrating their patient and business management applications into various wireless technologies to improve patient care at the scene and en route to the hospital. These efforts should mesh nicely with broadband.

How can the technology increase first responders' access to knowledge on the Internet and elsewhere that helps them provide better medical assistance in the field? Can emergency personnel use the network to tap into city or county information such as layouts of large building complexes, or details on hazardous materials which can be relayed to hospital personnel to bring their expertise to bear? What will the government need to do to make its data easier to access while ensuring data security?

Evaluate the various options to push emergency-related information through the network to the mobile devices of residents, care givers and public safety personnel. The U.S. Senate passed the Warning Alert and Response Network (WARN) Act, which funds an effort to create a unified nationwide emergency alert system. This bill provides funds for remote communities to buy technologies such as wireless to be part of this system if they don't have a good communication network in place, plus a general fund for others who can assist deploying technologies to support the systems. In addition to the federal government's efforts, cities and counties need to factor a local version of this type of application into their plans.

One application that doesn't get enough attention in the discussion of muni WiFi is embedded wireless devices that can go into assets such as infusion pumps, monitors, fire fighting equipment and the multitude of assets needed to deal with disasters. If you can use the network to help each public safety organization to track the location and movement of these assets, you immediately strengthen the security umbrella over your community. This also can put your community in line to secure government funds to offset costs of network deployment.

Economic development for the business community

The network infrastructure itself is going to be the linchpin to economic development in many parts of the country. Your municipality may already have done this, but an informal survey of businesses that have shown an interest in moving to your area should tell you how critical robust high speed communication is to attracting new industries. If they see this infrastructure as vital to their ability to ensure efficient operations and a competitive advantage, you should not only tout your efforts in this area, but also solicit their ideas for the network's enhancement. By making prospective businesses a "partner" in your network design, you can best align your network technology with their needs.

In some places, particularly in smaller municipalities and less populated counties, a broadband wireless infrastructure is needed to keep established businesses in the area that are main employers. Facing this situation, and after weighing the financial impact on the community of that business leaving, cities need to work directly with the company to determine its need and structure a wireless deployment plan that retains the company while benefiting the community at large.

To enhance economic development of mid-size and small businesses in low-income communities, a lot of what you're going to need initially is education which possibly can be facilitated by some of your partners. According to Loebig there are 200 businesses in this Olney area of one and a half square miles, many of them sole proprietorship and row house business. "Quite a few don't have any kind of Internet access because of the costs. There are many new immigrants and they haven't thought about it that much. Their decision to use it will depend on knowing what they can do with it. They need to understand the possibilities. Our mission is to show them how to do market research, manage their business and sell their stuff online. Very few have Web sites, probably because they don't have access to Net."

Community-oriented portals should be helpful in driving foot and online traffic to these merchants, but content and links made possible (though not necessarily developed) by the city's efforts is the catalyst for business improvement by even mid-size companies.

At some level, a city should plan to target business resources that in turn can deliver what communities need. DeCarlo observes "it's hard for us to figure out what are the right technologies to use because that means we have to spend time and money in figuring this stuff out, in teaching people, and then raising money for those things we need. I would like to have a page with two columns. In the 'A' column is a list from the experts saying you can do this, this and this. Then we show this column to the neighborhood and in the 'B' column they say out of that list, we'd like to do these projects and develop this service, or whatever."

To determine what technologies, content and online tools can impact businesses of all sizes throughout the city, the local chambers, tourism boards and local chapters of business associations are the likely partners to help you define these constituents' needs. They likely will explore ways to drive foot traffic and online customers to these businesses. Portals such as The Cloud (see Chapter 9) that was part of one downtown pilot project are important in this effort. They provide a variety of engaging content for residents and visitors while giving area retailers exposure and promotion opportunities. These tools also give businesses the means to create their own solutions.

To fully align technology initiatives with cities' and private interests' economic needs requires long-term planning and patience. Neither Rome nor great communication infrastructures were built in a day.

Bridging the digital divide

Broadband wireless offers city governments a potent tool to move underprivileged people to a higher standard of living. This is a major driving force for Wireless Philadelphia. The People's Emergency Center, the Norris Square Civic Association and Lutheran Children and Family Services were the first neighborhood groups to play a role in the project's early days.

Often the discussions about closing the divide center around giving people access to the Net and it ends there as if handing people technology is going to miraculously raise their economic status. This isn't a problem you solve with a drop-it-off-and-leave solution. If you look at Dianah's big picture approach, Wireless Philadelphia plans to get 10,000 refurbished computers into the hands of people who currently do not own one so they're ready for the rest of the solution which comes from the community groups. It's this program support and partnership development work that closes the divide.

There are additional elements you must add to the equation if you want to mount a meaningful attack on the digital divide. The big question is who are you going to partner with to fill in these blanks with online content and tools? Do they have a track record of success? Do these partners have the proven technology skills to provide what's needed, can they hire people with the skills or will the city's vendors help with this?

As partners define their plans, and you establish with them their respective responsibilities, be sure your IT people are aware of the partners' technology issues in case some part of the citywide infrastructure needs to be adjusted to address them. Define the role that social services, economic development or other city departments and their Web sites will play in the communities' overall plan.

An important side note here. I believe there should be some charge to citizens for the services that evolve from this wireless initiative, even if the charge for hardware and wireless connection is just a few dollars a month. The goal here is to close the digital divide so individuals can take more control of improving their economic condition and advancing their lives. You only foster dependency of a different sort if you give people everything for free.

Improving the quality of education

The School District of Philadelphia was ahead of the city in the deployment of wireless, and had implemented some aggressive Web-based programs to improve learning in the public schools. The District has focused its efforts on teachers and students in the classroom. Wireless Philadelphia brings an added dimension that hadn't yet been implemented in the School District's game plan – parent participation. With broadband access from the home, the parents can actively participate using some of the same online learning tools.

If your school system hasn't moved into this area yet, you may find that when you do look at the technology that's available today to improve learning in schools, a major bandwidth upgrade and wireless is necessary. Since a substantial portion was fundable by the government through their eRate program, the decision for Philadelphia to move forward was a no brainer.

"If you look at a district such as ours, there is such a dearth of quality data about how children are performing," observes Pat Renzulli, CIO for the District. "It's really a nationwide problem. You can't decide how to put together educational programs if you're not able to get feedback about students' performance in individual programs. This is a huge part of what our technology strategy is all about. I have three portfolios of projects."

What's needed are state of art classrooms, data-driven decision making and efficient business operations. Classrooms not only need enough computers, but also software to teach children how to build their technology skills and the software that enables individualized instruction for reading, math, biology and even for dyslectic students. These tools test children routinely and deliver online materials that are increasingly complex as a child becomes more proficient. Schools also need an instruction management system and benchmark testing that allow teacher to get feedback about students' performance and respond to that feedback immediately with appropriate curriculum.

In addition to technology that helps students and teachers, school districts in many cities need to manage huge business operations. Philadelphia's has the 2nd largest transportation system in the state and they need logistics software to optimize bus routes. The District's lunch operation serves up to up to 200,000 students a day and requires software to track sales and payroll. These operations require a huge technology infrastructure. Your school district may have similar requirements even if on a smaller scale than Philly's.

Moving beyond the classroom

The Wireless Philadelphia project came along in time to address what might have become the fourth portfolio deployment project for Renzulli. "We see parents as being next generation of beneficiaries for the materials that we are able to deliver," she says. "This instructional management system that gives feedback based on each child's performance also can generate materials for parents to use to help their children at home." The muni WiFi system, together with Wireless Philadelphia's drive to get computers for low income households, means the District can advance to this next level.

Melissa Long is Director of the People's Emergency Center (PEC) which participated in one of the pilot projects also sees potential to impact children's lives outside of the classroom. "PEC provides technology training for teens in our after-school program that improves their computing skills and ability to refurbish computers," states Long. "We didn't realize what an impact this would have and what opportunities we were opening up for them. Many of these students are always here, sometimes late at night, always wanting to learn more. A good percentage are going on to college for tech-related majors. Some of them become trainers in our adult programs and provide tech support to people in the community.

Teen programs such as PEC's address what is one of the great challenges of attacking the digital divide. You provide access to the Net and the various resources thereon. You develop or facilitate programs for providing access to computers and related equipment. But then who is going to train these new citizens of the digital world? Where do they find support for technical questions and breakdowns? If a school or community-based technology education program for teenagers take hold, having these teens become the equivalent of the corporate IT staff for neighborhoods provides a good practical education. Teens also learn valuable lessons in giving back to the community and communities learn a valuable lesson in creating sustainable self-help business models. Not a bad deal when you can make it happen.

Adult learning programs and skills development are issues that sometimes get overlooked in discussions on education. But if improving learning for all of its citizens is a priority for a city, then broadband wireless should address these adult issues. A city will need to identify groups that can play a role in delivering this service and work with them to assess if specific factors need to be addressed in the deployment plan.

Integrating with the collegiate world

Among university constituencies, cities will find that most, if not all, of the universities have deployed WiFi heavily on campus or have pilot projects in the works. Many of the businesses near campuses and catering to the college crowd likewise have set up hotspots. Your initiative is probably viewed by colleges as extending WiFi's reach throughout the city so students can stay connected from wherever they travel.

Municipal broadband may become an element of cities' relationship building with the college communities. For example, a city may work with colleges to ensure that the build out strengthens or enables the campus WiFi deployment. Or the campus network could be a testing ground for helping the city better understand its needs, the technologies involved, etc.

"Temple is part of the city, so it's in our interest for local neighborhood to do well," states Dr. Munir Mandviwalla. "Maybe this will act as a catalyst to improve neighborhoods and make them more livable. Also, who knows what new things might come out of expanding the boundaries of the campus? Though no one can envision some of them, change will happen because the technology is available. Why do students need to have their group meetings on campus? Now they can have meetings anywhere.

There is hot trend among colleges that may be impacted by muni WiFi, which is that many kids only want to go to urban schools and live on or near campus. All these schools that students shunned in the 70s now have a huge housing crunch.

"Temple is building dorms as fast as we can and working with private developers, but we can't keep up with demand. As a student, especially in the School of Business, you cannot exist unless you have access to a course management app called Blackboard. Often that access is through the Net, and students expect the university to provide it. So, to address both needs, you can go in and re-wire all kinds of old buildings, or you can piggyback on top of Philadelphia Wireless. Then you can buy buildings for students that don't have to be that close to campus, but those where you get the best deals."

Both colleges and cities should heavily promote municipal broadband as a draw for new students. The top question from many of the high school seniors visiting potential colleges is "does the campus have WiFi?" Since cities would do well to keep a lot of this college-educated talent around after they graduate, they should also consider how they can use content accessed through broadband to keep grads connected. Swartz, a long time resident of the city, believes a focus on collegiate constituents has long-term value.

"We have a lot of colleges and universities here. But the difference between here and Boston is that when students graduate from colleges like Harvard or MIT, they stay inthe area. When they graduate from U of P or the liberal arts schools like Haverford, they tend to leave because Philadelphia is historically a blue collar town. Aside from medicine and law, we really haven't built a white collar economy. The future certainly is in the Internet and using broadband access for national and international marketing. By creating a wireless city, it telegraphs that this is a frame-work for the city that we're all going to support. The emerging young professionals who are now using high speed for their laptops and PDAs will see Philadelphia as a place where these skills are an advantage."

Some universities offer students an e-mail account for life. Cities should ex-plore how grads can be sold access accounts for life. Wireless Philadelphia has looked at getting the universities to turn over management of student and faculty accounts to its service provider which may nor may not be feasible in the final analysis. The colleges would still control content and programs users create and access, but they wouldn't have to deal with the maintenance or administrative headaches.

To sum it up

There are so many elements to matching the right applications to the various needs that it may be difficult to decide which direction to move first. The key is to always return to the main objectives of the initiative, prioritize based on where you can get some early wins to build momentum and keep the feedback flowing.

After exploring ways to align technology with needs, the test of whether you made the right decision comes through a series of pilot projects. The next chapter gives some details on how you can maximize their use.

Chapter 9

Take Me to the Pilot Project

The pilot project is the process of taking a subset of the total mobile workforce or geographical area an organization plans to wireless-enable and equipping it with the technology in mind for full deployment. Pilots test the ability of the products to perform as advertised and withstand the rigors of the work environment in which they will operate, as well as the validity of the assumptions about the potential ROI that deploying the technology will generate. Sometimes the technical capability and ROI assumptions are tested at the same time, while other organizations only test one because they are pretty sure of either the capabilities or the assumptions.

In the last chapter I divided wireless initiatives into three categories: 1) the physical municipal network or infrastructure, 2) the mobile workforce and wireless-enabled assets and 3) the various constituent groups including businesses, neighborhoods, tourists, etc.

In regards to the network, pilots are mainly to test technology performance. Pilot projects for the various department applications test for performance and/or ability to meet ROI and business-impact expectations. These are similar in form and format with pilots at commercial entities. Finally, there is the collection of pilots for community and constituent applications and programs, but the format for these and the criteria by which they are judged is likely to be all over the map.

In all three categories, the business case is not always cut and dry. The value of government and community efforts to improve quality of life issues is not easily quantifiable, yet somehow you have to reconcile the fact that taxpayers demand some level of accountability.

As you develop your constituent pilot projects, realize that there are few technology or business performance benchmarks from other cities to compare results against. In a year or so this won't be the case. What's more, there aren't benchmarks for measuring the success of the business model. If the city/provider partnership is going to sell subscriptions to the network service, for example, there are limited parts of this business operation you can pilot test to gage ROI. Subsequently you have to be flexible in planning these, and open minded on how you evaluate the success of these pilots

Reviewing how Wireless Philadelphia and its partners managed their pilots provide some general guidelines for approaching some of these unique issues. Luckily, in the area of mobile workforce applications there are a good number of implementations that have been launched in all types of government and commercial organizations, so here you can find many more benchmark reference points.

Pilot projects 101

In a general sense, pilot projects are merely a dress rehearsal for the rollout of a major application. A lot of the process involves technology testing by the IT staff who get eyeballs-deep into the bits and bytes of a software program or the chipsets and screen displays of the hardware. They're trying to make sure each component works well on its own and integrated with other applications.

As an administrator or director on the business side, the primary questions you want the pilot to answer are 1) can people use the technology effectively, 2) what will it cost to get it deployed and have people use it effectively and 3) will the benefits you expect be the benefits you get? These three questions apply for the municipal network, workforce applications and community or constituent programs. The mechanics and logistics for determining the answers will vary widely for pilots in each category.

The municipal network

The infrastructure components for the network – the edge, the middle tier and the distribution tier - are typically first tested through what's referred to as a "proof of concept" in which a square mile or less is built out with all of the component parts. Then people are informed that they can stop by with their mobile devices and see how the network operates. In Philly, the project team did a proof of concept initially in Love Park and Earthlink will have to do one covering a larger area before full deployment begins.

Though the line between a proof of concept (POC) and a pilot seem to blur in these municipal projects, the former is primarily about showing that the technology can indeed work and pilots show to what extent the technology will work. The difference is similar to doing a test drive around the block a few times versus driving a car across the state over different terrains while testing all of the features and alternating between hauling passengers, cargo and a trailer.

Besides making everyone comfortable with the vendors involved, a POC is a good way for the business side to get constituents to understand what you plan to put in place and recruit supporters by giving them a sense of the benefits. The deployment process for the pilot gives you some initial insight to the time and logistics that will be involved rolling the infrastructure out to the entire city or county.

For the IT side, the pilots primarily consist of running extensive performance tests to understand topography issues, how the network handles huge numbers of users, the impact of climate changes and other technical features. These test results give the business side a better idea of final network deployment costs and operational issues such as equipment requirements and coverage limitations that the workforce and communities need to address when their pilots launch. "Establish up front clear criteria on what you intend to study and what you want to get out of the pilot," says Cole Reinwand of Earthlink. "An approach such as 'come over, put the network up and we'll see how it performs' won't cut it. You must be specific about what your needs are. What are you ultimately trying to do? Set a fairly strict timeline for the pilot to establish these criteria."

Raymond St-Jean, Manager, Integration Section Mobile Communications Services for the Royal Canadian Mounted Police, looks at the technology and what it is they're evaluating, and then tries to include as many end users as possible within reason. The objective of the pilot is to determine if this technology meets their requirements, is it stable, are they getting value from the product and what are the support and post-pilot issues, if any.

"I typically pick a small number of vendors and put them in environments where they can be compared apples to apples. For example, when we evaluated several 3G networks, we picked a region small enough to be manageable so we could see how all of the providers dealt with the same coverage issues. Some regions are more eager to be guinea pigs than others, so we like to choose those that will put in extra time working with the technology. We deployed the hardware, set everything up and let it run for a while. After we fixed whatever problems we found, we took the systems to a steady state so we could do quantitative analysis. Then we addressed any additional problems one by one and finally made our selection."

John Dolmetsch is President of Business Information Group, a systems integrator working on the Philly project, and he cautions that "you need to have a well thought out plan regarding the city assets where you expect vendors to mount the equipment. In different cities, access to the light poles has been a problem. It wasn't a forethought, it was an afterthought because cities thought they owned the rights to these and they even told us that they did. We've started pilots and then found out that one of the local utilities in fact owns them, so the cities had to go through various procedures to gain access. Make sure before you bring in a vendor that you can use different mountings whether on buildings or poles. If you can't, approach the owners first and negotiate all those issues."

To that end, a city needs at least one staff member to be the dedicated point person who interacts with the vendors. In the pilot stage some city departments won't understand what you're trying to do, so it's very difficult to expect the vendors to go to each department and ask for access to the poles, terminals and other resources. Along with a single point of contact, have an open process for gathering feedback from vendors throughout the pilot.

Workforce wireless applications

After being clear with IT about the mechanics of how workers perform their jobs and what benefits are desired from wireless, the next major role of directors and managers is to define the parameters of the pilot. How many participants? Will the group be representative of their department's various job assignments and skill sets? Are we testing in different work environments such as indoor, outdoor, hazardous terrain, etc.? How long will the pilot run? What benchmarks will measure success and can you accurately extrapolate overhead costs or benefits to the whole workforce?

You need to prepare for business interruptions since pilot participants require training in using mobile devices and possibly the new software. There's downtime while mobile workers and physical assets are equipped with devices. IT people who provide end user support for other applications could be diverted to mobile workers during the pilot's initial days.

A big challenge is measuring the improvements in performance and cost savings, so be sure pilots run long enough. There is some amount of lag time before you see results, and productivity might even drop while workers get up to speed with the technology. Tracking the metrics such as time saved or eliminated on a process can be tedious, so as much as possible have applications provide their own audit trails, such as reports that show how many clients were visited, violations processed and so forth.

Facilitating open communication is vital not only for getting valuable feedback, but also for using the pilots to build employee support for the application and quickly resolving issue that could cause worker push back when it's time for full deployment. The organizations that get the best results from pilots use an iterative process to get input before a pilot launch, get more input after participants use the application and then before the pilot ends, show participants the changes made based on their feedback.

Constituent projects

When you deal with constituent projects, your primary objectives are to show successful uses of the network, determine how well the infrastructure supports these types of projects and create a blueprint for other groups to follow. You probably won't be as involved with a lot of constituent projects once the network is fully deployed, but in the early stages you want to ensure that you create a momentum for the changes you're trying to affect in the city or county.

For maybe the next year there will be as many ways to run a pilot as there are constituent groups. This creates challenges in the beginning, but as you run focus groups and town meetings, the types of projects that constituents really want to develop bubble up to the top of priority lists.

As for setting criteria to test for in these pilots, refer to your main goals for the citywide network. If the main objective is economic development, then set up pilot projects that draw community businesses onto the network or drive customers to these businesses. If the objective is to increase a positive view of your city or region, then help develop pilots that reach out to tourists and visiting business travelers.

It is vitally important to track pilot results. How many neighborhood residents logged onto the Internet for the first time? What was the impact in terms of people using the network to find jobs, get responses from City Hall or other quantifiable activities? How many stores used the commerce-generating portal and what was the impact on retail sales? Quantify the number of students receiving technology training who are answering tech support questions from residents.

These pilots are where you can push your creativity to the limit, though you do want to limit the number of pilot projects while the wrinkles get worked out. Besides providing the early successes that increase support for full deployment, also use the pilots to fine-tune the list of technology issues that need to be addressed in other constituency groups and ferret out potential roadblocks to end user acceptance.

"However, I would offer a word of caution," says Scott Shamp, Director of the Grady College New Media Institute at the University of Georgia and driving force behind Athens, GA's downtown WiFi deployment. "Some end users unfortunately develop expectations which can be divergent from the intended purpose of the wireless initiative. If a constituent tells you that your system doesn't do 'X' when you really designed the system to do 'Y', then it is important that you put your feedback into the proper context for them. Don't let criticism distract you from what you want to accomplish with your project." At the same time, always be mindful of that you have to constantly manage expectations, even when people support your initiatives.

Philadelphia's pilot projects

After launching its proof of concept network in 2004 across the street from City Hall in a small area called Love Park, the city then put five official pilot projects in operation over the months that followed. The physical infrastructure was provided through donations of equipment from various vendors, including Tropos Networks, HP, BelAir Networks, Cisco and Lucent. There were actually several more pilots that started, but they were abandoned when the vendors involved decided to pull out.

Making the infrastructure work

Project Manager Varinia Robinson developed a document that described what the team expected from a pilot, including adoption of an area for one year, technical support and equipment that was all donated by key vendors. There were some moneys spent at Ben Franklin and Love Park but that was because this pilot initially had to prove the idea to the Mayor. The PEC network in Powelton that Cisco donated equipment for is not open to the public, but an open network is being constructed. The document also provided information for constituent groups about putting processes and programs in place.

Veronica notes. "Some of the vendors were bringing up areas while the RFP was issued. But I kept emphasizing that one didn't have anything to do with the other. The pilots were separate from the RFP. Several vendors decided to back out of the bidding and so they stopped their participation in some areas that we had already publicized. All through the project there were some people who said it can't be done, that the plan's not solid. So we said there are others who are interested."

In looking back over the whole effort, Dianah believes "the pilot phase helped us see how well certain companies worked together. Even with 12 responses to the RFP, only four came in as independent solo-company bids. Each of the other eight turnkey proposals had three-to-six companies coming together. You also learned what level of monitoring is needed, how quickly you'll be able to deploy, how to deal with signal interference, although we didn't have this with most pilots. Basically, you learn what it takes to maintain and support the network."

Cities have to look at a core of reputable technology providers during their pilots. You shouldn't limit your review to a single vendor or necessarily to a single type of technology. WiFi might not be the right solution if a city looking at video surveillance, for example. Cities need to be a little more open on the technology. They should describe the desired application, but leave it to vendors to propose and justify what technology best serves that need.

Don't ignore existing infrastructure when setting up your pilots. I read one consultant's financial assessment in which he recommended replacing existing high speed lines with WiFi because it would be cheaper than paying those recurring service charges. Cheaper, yes. But is it prudent? Morrow County kept their dual-fiber-lines infrastructure in place and integrated WiFi so they have super-high speed backhaul for the mesh (WiFi always has to have a land line somewhere in the network) and redundancy every which way you look. As much as you can, take advantage of complementary technology before actually replacing it.

Life's little challenges

"On the community side where we did portals, we learned that it takes lots of feet on the street meeting with groups, explaining what the program is, showing them how this can help them meet their goals," Dianah continues. "It was more than we thought would be necessary. How do you get these players who don't usually work with each other working together? You need to have multiple projects going on so you can move forward in some areas while waiting for others. Some areas just take longer to follow through on their plans."

I mentioned earlier the importance of gathering feedback from the mobile workers and constituents during pilots. It's also important to collect and analyze feedback from the network. Civitium's Greg Richardson reports that "the Love Park network provides volumes of output such as details on the way security is handled and how many people come and use it. We could have done a better job with the other pilots, but with different companies running those it was awkward to get information from them. I hope that when Earthlink does their Proof of Concept, they'll reach out and share this type of data."

The pilot projects underscored the struggles that come with acquiring sites for placing infrastructure equipment because finding rooftops is very difficult. According to Richardson, "You not only need the right location you also need the right vertical height, and the buildings' owners must be willing to grant you rights to put gateways there." In one of Philly's pilots there may be 15 Troppos access points per sq mile, and only three gateways to the backhaul. That's a 5-to-1 capacity layer or "backhaul injection." Earthlink is considering having 30 nodes per sq mile and 10 gateways for a 3-to-1 capacity layer. Though it means better coverage, the additional gateways also require a lot of work to find and negotiate rights to the additional rooftops.

Several articles have pointed to payment processing for low income individuals as a potential issue and one that wasn't addressed in the initial pilots since network access was free. In the next POC phase, Earthlink will no doubt tackle the issues because this is a market that doesn't have a lot of people with credit cards or checking accounts. Prepaid service cards are an option, but an infrastructure needs to be put in place to handle this.

Smaller towns wanting to pilot test a range of product options have the worry that they might be ignored by larger vendors with broader product lines because small installations aren't financially rewarding for vendors. However Shamp says that "the biggest lesson I learned from this process was the value of working with a vendor who understands and can accommodate your specific implementation challenges. Don't work with a vendor that tries to wing it by using what it already makes if those products don't meet your needs." If you have to wait for vendors to pay attention to you, look for alternatives or be patient. Don't settle for inadequate products and be leery of vendors who aren't used to working with your type of community.

License & Inspection's pilot project

Because of Mayor Street's Neighborhood Transformation Initiative (NTI) that was launched during his first administration, L & I knew that the city would be demolishing all of its abandoned and blighted buildings. One of L & I's roles in NTI was to have its Contractual Services unit provide lists of addresses of these buildings and create code enforcement case paperwork. Sometimes inspectors went out and looked over areas in the target area to make sure the unit didn't miss any buildings.

James Weiss reflects that "this project was increasing the intensity and volume of our activities and we knew we couldn't keep up using our old mode of operation which was paper based. We were already using code enforcement software from Hansen, so we got approval for funding to pilot test the application's mobile module."

L & I decided to have all nine inspectors in the Contractual Services unit participate in the field portion of the pilot, which was a good choice. It was a manageable number of people, they were handling the most pressing inspection tasks, particularly in light of NTI, and managers could get enough usage-analysis data so they could apply their findings to all 137 L & I inspectors.

There isn't a magic formula for how many people you need in a pilot. It really has to be determined by the nature of the application, what it is you're trying to evaluate and the size of the entire workforce. Discussion with the vendor after you complete your needs analysis helps determine what makes sense. There is a point however, when you can have too many people in a pilot for one application. Even if you have several thousand users, a pilot with more than 150 people could be unwieldy and counterproductive.

Depending on the number of participants, you can test several brands of mobile devices at once (maybe 10 people use one device, 10 others use a second) or sequentially where everyone uses one device for a time, then switch to a second and so on. L & I is doing sequential testing that includes a greater number of inspectors beyond the original group.

"When we first started the pilot, we were using PDAs," James says. "They are handy but the screens are too small to show everything that inspectors need to see, there's too little memory and the processor is too slow. Right now about half of the inspectors are using Panasonic Toughbooks, but we're looking at other devices for the other half of the team later and will probably go with tablet PCs. Conventional notebooks are too awkward in the field for most people to use."

The pilot testing for the software that the office staff will run was conducted in tandem with the field pilot. While the vendor was customizing the software, L & I set up a lab where representatives of the workforce who actually would be using the application tested it.

"Managers of the units selected the participants," continues Weiss, though some mangers might have tended to use only their best people. We made sure we had a cross section of employees so we would know about everyone's concerns. You know how there's always that go-to person in the unit who knows everything? We wanted that person. We also wanted the people who didn't propose to know everything, but knew certain things and did them well."

Whoever started with the pilot stayed with it. Managers might have added a person here or there, but they kept the original people. This approach ensures that the people providing feedback and reviewing updates to the application maintain a continuous view of the process.

One of the pilot's tests was to create failure scenarios such as trying to make the license inspection process fail due to a missing a social security number. The user would try to complete the business process without the information. If the software rejected the attempt, then the tester knew it was working correctly. According to Commissioner Robert Solvibile, "in the old system we'd create a corrupted data base because there was no data entry check like that. In this system we're trying to keep the database as pure as the driven snow."

Taking it live

After this testing was all done, the project moved to a Go Live stage. Every day for two weeks the team ran the system full-on in a controlled environment and had the users highlight errors. Jim tracked these and held daily meetings with the users. At the end of two weeks, as complaints were tailing off and things were running ok, the project moved into a less intense testing period."

L & I is implementing features of the software to address a set of business processes for each of their units. After each implementation's final "Go Live" testing phase everyone comes together to discuss what was done right and what they need to improve. One of the things the team decided was to have more than two users at a time for this phase since workers would get burnt out doing the testing while also dealing with their regular job responsibilities.

You have to be careful that you don't overwhelm your workers in pilot projects by having them do too much or try to complete things in a super compressed period of time. If you're doing the pilot right, there's a period when the new application runs in parallel with your current applications, which can mean having pilot participants doing double work. Also, people are writing up their observations for reports, which is yet more work.

As you move from pilot to full deployment, there comes a time when you have to close off the application updating process, take what you have and put it out there for people to use. Otherwise, you'll never get anything completed since you'll always be interrupting the deployment process and individuals' learning efforts to add "just one more thing."

One option to consider while you're in final test mode with office workers is to have your project team and the vendor's development team go out in the field for a day or two with the mobile workers who are going to use the application. I highly recommend that you also do this on the front end of the pilot process, but if your project team is too overwhelmed with work then at least put them in the field at this critical juncture.

Raymond states "I adhere to the philosophy that the project team has to be on site with users for some period of time. Nothing generates more animosity than to have someone from headquarters come in, install a system and then leave. The pilot won't be a success. If you can, set up one-on-one meetings between your project team and users. Ride along and deal with people as they do their jobs. This produces more honest feedback than paper surveys. You'll get a better feeling for whether or not the technology is working."

Mini roundtable discussionon pilots

Here are some additional perspectives on managing workers through the pilot project. This snippet from my book "Pilots to Profits" gives you insights from people in the public safety and health care worlds. Dr. Joyce Copeland, Medical Director, Community and Family Medicine, Physician Assistant Education at Duke University Medical Center, and Fire Captain Greg George and Electronic Documentation Coordinator John Pringle with the San Diego Fire and Rescue Department, deployed mobile data collection applications.

If an organization has two camps with strong but differing preferences for mobile devices, what steps do you recommend to resolve the situation?

George/Pringle: Quite a few people were familiar with Windows, so some wanted Pocket PC devices and others wanted laptops. However, it all came down to the size of the budget that dictated we go with the palmOne devices. The laptops and Pocket PCs were more expensive. palmOne also had a lot of companies making software development tools and they were relatively inexpensive for a pilot.

A lot of people hadn't used these small devices so as the pilot progressed, the most common question was 'when are we going to a laptop?' But both of us are former fire fighters, and we felt that the smaller devices were best suited to the needs in the field, which made it easier to sell this option within the department. We developed the best end user interface so people can operate the same as they always have, but faster and easier. After using the palmOne units, people didn't want to use laptops. If you have resistance and you have the finances, try giving users multiple vendor options.

Copeland: This wasn't an issue for us because we didn't give students a choice. But over in our Medical Center they had trouble because of the debate about Pocket PC-based and Palm-based applications. The physicians had been using Palm devices because initially Palm had more medical-related apps. IT favored Pocket PCs, and bought them for the whole staff. No one was talking to anybody to get a consensus. So what ended up happening is that they have to support both devices that access the same application, and it's a lot more work.

To prevent these kinds of problems you have to get to workers early before they make decisions. If you have people who are already using devices, someone will be ticked with whatever decision you make and they're going to fight you. So make the decision that's best for the organization, communicate it to workers and try to get everyone to move on.

2. How did you select people to participate in the pilot?

George/Pringle: Geography was the criteria since we wanted a place somewhat isolated, but also containing first-responder units and EMS vehicles. We chose a border area with Mexico because it encompassed the department demographics as well as the general dynamics of the city (employees nearing retirement in some stations, young bucks in others), in a small geographic space that was easier to manage physically. In the end, we were very please that we selected both more recent and veteran employees because the results of the full deployment were similar to those of the pilot.

Copeland: In the beginning everyone got palmOne devices 6 months before the application was developed so they could play with the devices and get comfortable with them. Then we set up a server with software and updates so all of them could download the application. Next year people will get it when they join the student program in the fall.

How do you recommend that managers select and manage people in order to get the best results, feedback, etc. from the pilot?

George/Pringle: Some managers think that for a pilot project they should only pick people who are their best, most productive employees because they will be more interested in the technology and better participants than less 'energetic' employees. We wanted both types because we knew that we had to make the application work with the least motivated employees as well as the gungho types.

Using a cross section of employees gave us a picture of what would happen, such as what steps we would need to take to protect the devices from damage in the field. The one thing we had to do was lay the law down that, unless there was a technical problem, they were absolutely required to use the devices.

Copeland: If I could do it over, I would run more pilots and do them with people who can give me educated feedback. By having everyone participate we were working with students ranging from technogeeks to technophobes. We didn't get a lot of good data back. Our nomenclature has a lot of synonyms and different people may call the same things different names. You have to come to conclusion about what you will call things. Make the application as graphically friendly as possible. Use as much point and click as possible. For details that are entered repetitively, have them automatically populate the fields.

Pilots in the community

The number of potential constituent projects can be dozens or in the hundreds for larger municipalities with ambitious social agendas for the technology. When you can, use the early pilots to create a foundation that communities can build on. However, as Dianah stated earlier, you can't use a cookie cutter approach. There's a balancing act here between not re-inventing the wheel and dealing with the uniqueness of your constituents.

This section looks at two constituency groups in Philly and how they approached the pilot process. There are a number of lessons you can pull from here to adapt to your particular circumstances.

Working with their heads in The Cloud

The history of The Cloud starts in Athens, GA during early 2004. A consulting firm named Pervasive Services helped the city with a project to leverage their municipal WiFi deployment to drive business to downtown merchants. Pervasive licensed the intellectual property for the resulting software from the University of Georgia.

During a meeting with the firm, Civitium felt the important attributes of what Pervasive is doing fit with what Wireless Philadelphia is about. "Cities that are doing municipal wireless deployments aren't looking to be an Internet provider," says Pervasive's Director of Marketing and Client Services Bailey White. "They want to do things to help the citizenry in tangible ways that represent the area."

The Cloud is software cities use to create a captive portal that people sign on to for Internet access. While they're online, the software allows users to connect to places right around them. This application is very much focused on the local attractions, and it is flexible enough to be adapted to different environments. "We were coming from a downtown environment in Athens that was rich with music venues, bars and restaurants. Philadelphia is different. Its downtown has several concrete parkways surrounded by a lot of green space and museums. It's very different in terms of social dynamics."

The Parkway nevertheless was an easy place for The Cloud to start in Philadelphia because the city was running this pilot project themselves. They could control the whole application and tailor it to meet city objectives. Other pilots were being run by vendors that had supplied equipment.

The value of this particular type of application is that it enables cities to provide that balance between giving constituents a foundation on which they can structure application development and the freedom to be unique. The software that drives the portal runs on top of the WiFi network. There's a publishing tool that local businesses and museums use to post information and promotions, with a special mobility feature that lets the software's manager push out information to people who sign on to the portal and give permission to be contacted.

Both project managers and local businesses benefit. The publishing tool prompts businesses where to drop information and everything gets uploaded automatically in the proper places. But it's more than that, Bailey states. "It's dynamic. Businesses get an account and whenever there's something happening they think is relevant, they submit it. We'll check it out. The item will go to the portal or out to mobile devices if we think it's relevant. A large benefit of municipal WiFi is that it allows people to be on the move and able to do a lot more from different places. Museums can send notices to people's cell phones to let them know about unique things that are happening. Tourists can look for events and send details to others about what's happening. This is a living publicity channel."

Content, which can be uploaded wirelessly or via a Web browser, is accessed from mobile devices or from desktop computers. The editor for Philly's Cloud during the pilot is an intern who lives in Philly and is managed by Philadelphia and Persuasive. In some cities, the firm has managed and edited The Cloud on its own. The pilot launched in April of 2005 and will last a year.

Live 8 puts The Cloud to the test

Philadelphia held a massive Live 8 concert on the Parkway July 2, 2005 that packed the area with tens of thousands of people who love to party in the sunshine.

Pervasive partnered with Greater Philadelphia Tourism Marketing Corporation and they worked together to promote The Cloud and Wireless Philadelphia three days before the event. This promotion resulted in 2000 page views a day over the weekend The Cloud Webcast events from the concert.

Journalists from XM Radio, the Boston Herald, New Jersey Star Ledger, Technorati and other outlets used the network throughout the day of Live 8 to update their media outlets.

The Cloud made concert goers aware of offerings in the city via WiFi, its Web site, and text messaging. Some of the results delivered during the weekend were simple, but appreciated, such as finding better places to sit using a Cloud text messaging notification, or locating the spots where firefighters were giving cool sprays to hot revelers.

Other results were financial. Denim, a popular night club, got extra traffic because The Cloud recommended their late-night Live 8 event. The Cloud offered a gift certificate to a local Parkway restaurant as part of a photo contest asking for images from LIVE 8 spectators. These were posted in real time along with stories and blogs about the event while the restaurant received a boost in customers.

Local students from Drexel, the Art Institute of Philadelphia, and recent graduates from Temple and Kutztown universities volunteered their time to help make The Cloud relevant for Live 8 attendees. They surveyed the network repeatedly to help tune its performance, wrote and posted articles to The Cloud comparing the Philadelphia of the original Live Aid concert to the city now, reviewed band line-ups, and promoted Wireless Philadelphia to local, student and personal email lists. Over 2.6 GB of data moved across the network during the long weekend.

Looking at results overall for these first months of the pilot, Bailey feels The Cloud has been very successful. "We have over 2000 members who created accounts, mostly locals but some tourists. We have done great outreach to talk about what Wireless Philadelphia is, and have worked with most of the museums, retailers and other businesses in the Parkway area. We get usage from some of the hotels."

With all of these pilots, groups are at the beginning stage so it's hard to measure the financial results. Most of the impact to date is in the interest and the media coverage generated as opposed to revenues generated. 2000 members are good numbers for six months, but nowhere near what Baily feels they can get once they start doing the full amount of outreach. What's in place now is great for building general awareness, getting people to understand and use The Cloud and inspiring organizations such as tourism groups to think about how we can use this tool across the city as it becomes wireless."

The neighborhoods are alive with the sound of WiFi

"We had no intent of trying to get WiFi when we heard the first stories about it being used in other cities because we didn't think it would be available in Philadelphia," recalls Brian Loebig of the Lutheran Family and Children's Center. "We already had a T1 line, but it's just used by the staff and we weren't making use of WiFi connected to that." Then Wireless Philadelphia came along and their North Philly Olney area became one of the neighborhoods chosen for a pilot project.

Loebig's organization manages a seniors program in one building in the neighborhood. There are 50 – 70 people a day using it now that the whole room has wireless access. Wireless Philadelphia made laptops available for use by Tech Access, our new non-profit startup in the building. Tech Access has set up these and additional systems for seniors to use.

These tech savvy seniors are accessing genealogy information, health-related content, and recipes. They weren't able to do much with the Internet before because there was only dial up access and so many Web pages today rely on dynamic content that you really can't use well with a slow connection. With the higher speeds the thrift shop posts photos of its inventory and sells furniture online, which normally would have required T1 or cable service.

Everyone involved with the Olney pilot agrees that a lot of its success is directly attributable to the huge neighborhood launch party they had to kick off the opening of the network. Karen Archer Perry was Lucent's community relations point person at the time and she felt there needed to be a major event to draw the community's attention to the project.

The project team brought out 20 giant patio umbrellas to 5th Street for the July event and under each one was a person or local business showing off something people can do on the Net. All of these demonstrations were WiFi-enabled. There were online games, an immigration attorney presenting his online services, someone else discussing health and fitness while pointing people to related content on the Web.

Karen says "we're dealing with a community that has just a basic level of computer adoption and Internet awareness, so we wanted to introduce both Internet technology and the wireless service while at the same time building the desire to be online. I tried to come up with a diverse set of presentations thinking that you might not like all 20, but if you saw two or three that were compelling to you, this would inspire you to get onto the Net."

The other element contributing to the event's success was a group of about 100 volunteers, many of whom were from the community, sharing something with their neighbors that they thought was important about being online. They were ambassadors for the project. They understood conceptually what this service would mean to everyone, and they understood the practicalities of how to get onto the Net. The event was a great opportunity for them to bring others aboard the program.

Following this event, Lutheran Family and Children's Service converted their nursery that wasn't getting much use into an Internet Café that offers community residents Net access and basic computer training. Every other Thursday while the weather was warm, three umbrellas were open for business in front of the Lutheran church and the organization's directors answered questions. Even without publicity, there were always crowds of people trying out the network and signing up for service as well as computer classes.

Brian reports that they're moving into phase two of the pilot, which is focused on increasing the integration of the network into the neighborhood. "We're continually signing up people, using remaining Lucent Foundation money and other resources to get computers, wireless cards and adapters for people, boosting the computer lab and providing training after Tech Access helps with installations. Publicity never stops."

Measurement is also a key element of phase two. Wireless Philadelphia and the project team track how many people use Internet, how many receive training, the amount of equipment that's distributed and other statistics. Their community portal from Ninth Wave Media helps this effort by requiring people to sign up for an account so the group knows how many unique individuals are joining.

However, one challenge during a pilot is that some community organizations don't have people with the expertise in using the tools to run a portal. You may have to work with a vendor that will link their portal software to the organization's Web site and manage the entire portal operation for them. This is similar to the ASP option for applications I described in Chapter 7.

"Survey your communities to see how much expertise is within their groups to run the portal," says Ninth Wave Media's President Judy Miller. "There are often one or two people who have some level of technology expertise, or might be willing to learn the basics. Assign a technology liaison from the city's project team to work with whoever is running the pilot. Then it's a matter of finding how in-depth the community project leader wants to go. How much do they want to learn about self-publishing tools? Some organizations just want to post a calendar and others want more features. We found all this out during the pilot and we're still learning."

As the uptake in the community increases, Lutheran Family is starting to see ways in which citywide broadband will benefit all of its mobile workers and help the organization overall improve its business operations. A lot of the information and computing resources used by the staff, including e-mail and their case management database system, is Web-based. With WiFi everywhere, everyone is free to work outside of their community centers and offices.

"Anticipating the point when WiFi is all over the city and allowing social workers to be able to access critical data from anywhere is giving us creative ideas on how we're going to deliver services, remarks Loebig. "We currently have to write down notes in the field and come back to the office to do the work. We run a big truancy program in Northeast Philly. When the network is citywide, we're going to experiment with having parent truancy officers use Palm devices to track case notes and access contact data."

To sum it up

The pilot project is a pivotal part of the any wireless deployment, whether you're dealing with building out the citywide network or neighborhood projects. This is the big test before you commit the big bucks to the technology you feel is going to make your initiatives a reality. In many cases, these pilots are your first chance to make a good impression, convert skeptics, strengthen your project champions within the mobile workforce and the constituencies and see how vendors' products work in the real world.

As Philadelphia moves into final contract negotiations with Earthlink for full deployment, I'm a little amused – and a little concerned - by the flurry of announcements by various municipalities across the country about their deployment plans. There's an undercurrent of a competitive motive behind some of these announcements, the "we-gotta-be-first" tone.

Folks, we're talking about spending millions of dollars to transform how cities and counties do business, provide vital services, and transform neighborhoods. The pilot process is not a race or a mere item on a tick list (the failing of some pilots) to finish quickly so the mayor can have bragging rights at the next national conference. This is about whether you spend those millions appropriately to achieve the main business and social objectives. The theme of every release should be "we're doing muni WiFi and we're going to do it right."

Take your time with both the planning and the execution of your pilots. Many of the Philly pilots are slated to run a year, though some of the specific constituent activities might be events that run for limited times such as those for Live 8. Gather good feedback before, during and after your pilots. And if the pilot shows that some assumptions are a little off target, or a product or technology isn't quite fitting with how the workforce or community works, go back to the drawing board and do a new pilot. This can be the hardest thing for a project team to do because muni WiFi is a very public and an incredibly political project right now.

"You have to look at how fast you jump into a technology that can potentially change how people work," advises Howard White, Deputy Building Official for St. John's County, Florida. "You shouldn't promise too much before it's been thoroughly tested. If you make expectations too high, then you can have a problem when reality sets in. Even with the newest application we're getting ready to release, which is really just a new feature we're adding to our current software, we have spent eight months making sure it will work before jumping out there."

From the beginning set the right expectations for everyone about how much time this process can take, and remind people often throughout the pilot what it is you're trying to achieve. As the President of mobile ASP ServiceHub told me in an interview several years ago, "what you're buying with these pilot projects aren't your first steps of automation, but useful information to make better decisions. So run multiple pilots. Don't be afraid to pay your tuition for your education."

Chapter 10

Moving to Full Deployment

There comes a point when those driving wireless initiatives need to conclude that they've done sufficient planning and testing and it's time to move forward. Whether people realize when that point arrives is another story, given that they can always find something more that to add, review, guard against, etc.

As Wireless Philadelphia Board member Robert Bright sums it up, "without question, the more input you get when identifying constituents, customers, and so on, the better. You cannot ignore the political world given the lobbyists and incumbent businesses. You need to get a good understanding of what are the potential obstacles out there. But when I put my private industry hat on and look at this, I have to make decisions. I can't afford to circle on all of my decisions. Government will not move forward by continual circling.

When you ask "does this address everyone's needs?" "Is it good for 'x' or 'y' organization?' "What if Verizon wants to do this or that?" and get all of these questions answered, you spend a lot of time. Do this too long and there's a point when you run out of gas. Robert remarks "we could still be doing the business plan today if I look at everything that's happened in the business world with this technology from last summer to this summer."

This chapter is about moving forward. It captures a great deal of feedback from vendors and service providers involved with the Philadelphia and other municipal projects. These individuals have years of experience overseeing deployment of large complex implementations leading-edge technologies. There's much to learn from them.

Engage all engines - make the RFP process work for you

Though your government may have developed RFPs for many technology deployments, few are similar to wireless implementations. Besides the wide range of computing devices and software you need to purchase, there are service providers, evolving network infrastructure issues, spectrum issues, business model options and community considerations you must address. This is not a job for the faint of heart and weak of planning skills.

When you start translating needs into an RFP, Cole Reinwand at Earthlink believes it's important to have the point person driving the RFP process do some filtering. A few cities ask all the departments what they need, what would they do with a wireless network if it were in place, and so forth. Then they put every one of these into one RFP and say a vendor must be able to provide all of these services. Prioritize your list and restrict it to the top five-to-ten things that would be most helpful. If you ask for everything and get it, this will become a watered down system or the RFP will be just too expensive to respond to.

Do thorough due diligence on technology requests from end users. If a supervisor thinks they need a gigabit-per-second connection, your person should say "you can't get that with WiFi. You're thinking about fiber and we're talking about wireless." You want these kinds of issues addressed before they make it into the RFP. Otherwise vendors are in a bad spot, either having to do a lot of time-consuming expensive education about these flaws, or committing to providing services that in reality they can't. It's hard for some vendors to say no because they feel that saying "yes" to everything is going to help them to win the bid.

This isn't your father's technology RFP

Rizwan Khaliq from IBM advises that your RFP reflects a broader strategic view than just that of your city departments. "Consider, for example, deployment of a public safety application for police and fire departments. If you look at a large area county such as Miami Dade, the Gulf Coast or some other areas that are exposed to natural disasters, a municipality that deploys a broadband network needs to get participation from multiple players. Those players are morethan what's under a municipality's umbrella, such as Federal or regional agencies that need to have access and be able to share data in real time with local officers. This is not a silo or vertical application."

What strategic view you have, and by default which other players you consider in the needs analysis, is different based on where you are and what natural disasters you're likely to experience. Another factor that might affect this view could be terrorism, the potential for attacks on historical sites or vital infrastructure such as a bridge in a town that impacts the ability to move critical resources to where they're needed.

Your strategic view should include your business communities. Rizwan continues, "a government can look at how companies located there could actually leverage the network for optimizing their business operations. For example, in Rhode Island look at how CVS, which has a strong distribution center there, is leveraging their municipality's WiFi technology. A business in your area may want to optimize the network to maximize fleet management." This can be wrapped into a mission to make the city more economically attractive to local businesses.

Commercial and public safety interests also overlap in this discussion. Emergency first responders may point out the need to be able to communicate with a company in an emergency if they have petroleum or chemical supplies on their premises. Having the license data from all buildings plus floor plans available to be sent to first responders, similar to what Commissioner Solvibile described in Chapter 3, can save time and lives. For a major area evacuation, these workers need to know what those business facilities look like before they get on the scene.

Is your point person in the know?

Given the mission critical nature of some of the city departments that access a broadband network such as this, consider this advice from John Dolmetsch, President of Business Information Group. "Your RFP should include similar requirements that have been in RFPs for deploying public safety wireless radio networks. There are established and documented procedures for the public sector that have been in place for 50 years now. Vendors need to respond to the RFP to build a WiFi network just as vendors would for these critical radio networks that have to be operational 24/7."

Every city, county and state has a public safety director. These are mostly wireless radio guys who rely on their push-to-talk systems every day. They already have the resources they can go tap to help you develop your document, including old RFPs you can review. "One of the problems we have in the industry is that in a lot of cities, the IT folks are running the WiFi show and they haven't involved the public safety director who has a lot more experience in wireless than the IT people do," says John. "It's possible the project person from the city may not know to point the vendors in this direction."

Another thing to consider is how much do the people driving the RFP process understand about the specifics of wireless technology and related issues. Are they willing to admit what they don't know? Tropos Networks was part of the vendor team with Earthlink that won the bid for the Wireless Philadelphia project. Their Marketing VP Bert Williams cautions that a lot of cities need to be realistic about this.

Dianah was practical in her approach. "We put together two panels of people when we reviewed the RFP responses, a business group and technology group. Civitium led the technology panel. They worked with representatives from Drexel University and University of Pennsylvania, private sector companies that were not involved in the bidding and independent engineers. They evaluated all of the proposals and came back with recommendations. I led the business panel. We looked at how well the responses met the business and social program needs, analyzed their costs proposals and did research on companies' financial stability."

If they have the skills internally to articulate the needs and write an RFP, that's great. But if they don't, then they need to recognize this and seek some third-party help. They can get in real trouble if they spec a system without really knowing what they're doing. They have one set of expectations of what the performance should be and the vendor has a different set. If the vendor feels they delivered to what they promised but the city's expectations haven't been met, then this obviously can create a serious problem.

As more deployments are launched, vendors are able to spot RFPs designed by individuals who don't have a good grip on the main issues, and one of two things may happen. When the size of the contract is enough or the vendor likes you, they will point out problem areas and help you re-work those. "If the city or county has way under-scoped the budget and they can't get what they want but they think they can, we'll definitely no-bid those kinds of situations," Bert warns. What often happens with an organization in this latter category is that they go through the process, look at bids that do come in and maybe even run some trials. When these don't go well, the project team pulls the RFP and goes back to the drawing board to re-write the specs, scrap the program or find more budget.

An upside to the increase in deployments around the country is that some elements of the RFP process that are successful in one city are showing up in other cities' RFPs. "Get a copy of Philadelphia's and Portland's RFP," advises Intel's Paul Butcher. "These cities did a very nice job of defining what they need and the process they want to follow to deploy the technology." These RFPs are iterative. Cities are plagiarizing each other as the latest and greatest comes out. They started by looking at the Philadelphia RFP and got better as they went along.

Paul does warn, however, that you don't want to use someone else's RFP to shortcut the process of defining your needs. Use them so you can figure out what you need to know and what you should be asking for from vendors. "Those who shorten the process end up confusing the private sector. There's one major city that has created an RFP that they did in a hurry and I hear there are vendors walking away from it saying 'we can't respond to this.' If you want a good partner, you have to do your homework."

Vendor roundtable offers deployment insights and guidelines

To give you an overview of important issues you should be considering, I brought severalleading vendors in the muni WiFi space into the discussion to share some insights. Included in the Q & A are:

Paul Butcher, Intel – Marketing Manager – State and Local Governments, America's Marketing Group

John Dolmetsch, Business Information Group – President

Rizwan Khaliq, IBM – Global Business Leader for Digital Communities

Craig Newman, Motorola – Director of Business Development for Motorola's Canopy Wireless Broadband business

Cole Reinwand, Earthlink – VP Municipal Product Strategy and Marketing

Bert Williams, Tropos Networks – Senior Director of Marketing

It's important to have this input because these providers of technology and services are very much involved with the wireless initiatives of municipalities worldwide, including some international cities that are much further along in their deployments. Their perspective is also valuable because they have a clearer view of the big picture issues, as well as the possible future direction of technology. There are some good tips here for selecting and working more effectively with vendors.

1. What do you feel is the role of a municipality and vendors/service providers in a partnership to bring this type of technology to citizens?

Reinwand: In larger cities, the role of government is first to be the champion of the project. There is a lot of political value and capital created by launching these types of networks. Whether the driving force is political or from the IT side, the municipality should build the excitement. Another thing a city brings to the partnership is rights to the assets where you can mount the radios – the traffic signals, street lights, etc. In some cities we're seeing RFPs where it turns out the city doesn't have the rights to these assets. It doesn't create much value to the partnership if the vendor ultimately has to go and negotiate with the utilities or whomever owns the assets. The cities have a lot more leverage with the utilities than vendors do.

Governments need to endorse the service with the community because this really helps the vendor drive their member acquisition costs down. You look at the brand equity of a city and you find that many citizens want to use the services that the city offers, they have an affinity for those services.

On the vendor side, their responsibility is to identify the correct technology, install it, keep it in good working order. Service providers need to drive users to the network, and also be open to alternative business customers, not just the consumers. They need to look at applications such as automated meter reading. It's essential that a network operator keeps their options open.

Earthlink is transitioning. It has historically been an Internet service provider and has relied on other broadband companies to provide the access through their networks. We now have the opportunity to move more into a service provider/ network operator role with municipal networks.

Newman: The role of municipalities depends on there agenda. If they're looking for a public safety network such as Oklahoma City is, or a public works network such as you have in Corpus Christi for automated meter reading, it might make sense for the City to run everything. What we have seen to date is that larger cities then want to offer public access. Due to market pressures and the bandwidth demands for running a public access network, it appears that this type of network is best run by private industry.

On the other hand, unless they are a suburb near a larger city, smaller cities and towns probably will need to build the network themselves as the capital markets have not fully embraced small private firms in this field yet.

Williams: To me the important thing is you have to structure a relationship that's going to benefit everybody. On one hand, we're interested in selling as much equipment as we can. On the other, we're in the business of having customers who are happy with what they get and are satisfied with the return they get on their investment because in the long term this is what helps us sell more and also have referenceable customers.

The main thing for the city to do is identify what their needs and requirements are. We're going to have different recommendations on how to construct the network if it's going to be used for public safety only, versus public safety plus city government-worker use, but no public access. Or public access only, such as residential and small business. Clearly articulate what your goals are so we do the best job of defining and designing a system to meet those needs.

Vendors need to understand that cities are not competitors with each other. With cities, it's much more collaborative. People who work in public safety or the CIO's office tend to talk to peers in other cities. In this market it's very important that vendors have satisfied customers. If they do, other prospective customers will hear about it. And if they don't, prospective customers will hear about that too.

This is not a one-time transaction. There is any amount of post sales and post installation support. These are radio networks, so the conditions under which they operate can change over time. Buildings are put up or torn down, trees grow or get cut down. There can be a need for ongoing maintenance of the network as well. We tend to remain in pretty close contact with customers even after the network is operational to make sure if performs to their satisfaction, and if it's not, to make the necessary adjustments.

Butcher: Equal rewards and risks is the only way that this can work. It seems to me that with the cable roll out, the cities gave up their rights completely. Right now they look at cable TV and see what they could have had. They gave up their seat at the table to have a say in price, who gets it, where it goes.

This time with wireless the people who felt like they got burned want to make sure when they negotiate these deals, they get a seat at the table for the long term. They want a role in deciding who gets it, where it gets deployed, how much gets charged and how often the technology gets refreshed. They realize that they have valuable assets in these light posts and roof tops. They have leverage and they need to build into their contracts these leverage points. But the cities also have to share in the risk somehow.

There's no financial equity when the city doesn't put up tax money, but there are other ways for providing value. One is alleviating the cost of mounting rights. It's like the farmer who negotiates a deal with a cellular carrier to receive payment to allow them to put a tower on his or her farm. A city is in a position where they can put a fair market value to its assets and put that up in lieu of bonds or tax money. This way, both parties have a stake in the project's success and both will lose something if it fails.

Dolmetsch: We're at the early stages of the game right now, but the free equipment from manufacturers for pilots is going to go away shortly. Everything is so labor intensive and the equipment is so expensive that the whole proof of concept phase such as Earthlink is doing will be waning shortly. Once you figure the equipment works in some other city, you can pretty much replicate that success anywhere you go.

Cities have to promote the network effectively. Before rolling these things out, they need to come up with a marketing plan. Who's their target audience, what's the message for constituents? Is encouraging mobile commerce for local business the real initiative? Philly did a good job at promotions.

2. Can vendors with a financial interest in the outcome be an objective partner in key municipality decisions?

Newman: Municipalities need to know what their priorities are, as this will dictate which vendor solutions and decisions are better than others, and minimize the issue of conflicting interests. Each vendor in this space has its strengths and weaknesses. Cities are experienced in dealing with technology vendors in general and have historical context to draw from. Nevertheless it would be wise for a city to employ a consultant with RF and Muni WiFi experience to help them in their decision-making process.

Reinwand: The most important thing to remember is that one can't operate without the other. It's very symbiotic. The cities are the ones that have the poles and other key assets. The network operators bring their technical expertise, whatever hardware deals they've struck and so forth.

Everyone needs to be in a mutually beneficial partnership. Cities have their interests, whether it's social objectives such as closing the digital divide, giving mobile workers wireless access to help them be more efficient, or even replacing expensive telecom infrastructure such as T1 with a newer cheaper technology. These are the kinds of motivations we're seeing from the cities and vendors need to respond to this.

Williams: This can be a double edged sword. As I already mentioned, we want to sell so we obviously are going to put our best foot forward. However, if we don't think we can make a customer satisfied, we'll decline to bid. It's also the city's responsibility to evaluate options objectively and be realistic in their expectations.

Dolmetsch: I think that sometimes vendors are not objective enough because they want to push products and technology. Whether that is providing too much infrastructure rather than thinking a little bit about of the best option, or trying to shoehorn a certain technology in where it doesn't fit. Maybe they don't consider the existing landline assets. If the city does have fiber, then use it as backhaul. Look at a city's entire current infrastructure to weigh re-using it in some shape or form.

Sometimes the best way is to ensure objectivity is to hire a consultant to identify your requirements. But to pick a good consultant, you are probably better off asking system integrators for references. Integrators have been around longer than most consultants and they work outside of the WiFi space to provide general wireless technology to public safety and other government organizations. The consultant can then work with the systems integrator to evaluate vendor proposals.

3. What are some tips to government organizations to help them work more effectively with vendors and service providers?

Butcher: Similar to how corporations do it, you first have to build in mechanisms that account for the fact that technology changes quickly. For one thing, you need a sustainable business model with revenue streams that allow you to be able to pay for upgrades and new technologies that come along. You can't do that with a free network.

Another thing is, you use the RFP process to acquire technology that's going to be around for awhile, or at least ensure it uses standards that will likely be viable in upcoming years. If you select vendors whose business operations and product roadmaps conform to your requirements, you should have a good working relationship

Dolmetsch: Make sure that the vendor or systems integrator has done this more than one time in their career. They need to understand the entire project. Just because it's wireless doesn't mean there aren't a whole lot of other technology elements – security, switching and routing, bandwidth control. Though it's difficult in these early days, it helps if whatever they've done has been operating for more than two years. Do they have proven statistics showing mean time to failure, system reliability, those types of things? Most of the vendors should have management systems that produces those stats and normally it's a requirement. It is in Philadelphia.

Refer to archived RFPs for public safety applications to see their requirements. They usually ask for audited financial statements and balance sheets. The vendor can't be two people working out of a garage. Do vendor visits. We won a build for the city of Wichita, and the city sent people to our offices to make sure we were real before awarding the contract. Checking out references is very important.

As far as managing the project, after getting the contract vendors should put up a performance bond. The insurer that backs the bond guarantees that the application will work, and they can kick the vendor off a job and bring in someone else in if it doesn't work. If the vendor isn't a solid business, either they won't be able to get a bond or they'll have to pay a lot of money for one. Usually a vendor's viability to get a bond is uncovered during the RFP process.

During the project you should have weekly status meetings to say 'where are you, what is your progress.' There should be documented network results and coverage maps.

Newman: Understand the financial solvency of each service provider. Be aware that some vendors will make claims about products that are not ready for market yet. If you're going to use the RFP process, make sure that the decision committee has good technical representation. It's advisable to bring in an experienced integrator, and make sure either you or they do the required homework on security requirements and processes.

There are some general things you can add to your check list for creating the RFP. Document your city assets in advance. Offer anchor tenancy opportunities such as Portland did in their RFP. Finally, have a first draft of the Franchise Agreement (or Contract) complete in advance.

Khaliq: Before awarding the contract, one of the key values that governments should look at is whether that partner or vendor understands the government industry, as well as understands the broadband industry. Often when a government group looks at a major technology application, they need someone who has a more comprehensive view of how all parties can benefit. This is where a systems integrator such as IBM or a consultant can add value by bringing the necessary parties together. in terms of understanding each other's needs.

You may get vendors that have a good grasp of issues regarding government operations, but don't know what the issues and concerns are that pertain to the network. You have infrastructure, software applications in different departments, mobile device issues, service provider issues, customer service. There are a lot of elements in a citywide deployment. Each vendor can tell their piece of the story but not put it in the bigger context.

Once you begin deployment and during the implementation, the government needs to be partners all the way through as far as managing the process is concerned. But they're not in business to be a network operating center (NOC). Government would be better served to focus on what they're in the business for. Likewise vendors need to do what they're best at delivering.

Williams: After you start deployment you have stay informed if sections of the city's workforce decides later that they want to use the network, though they weren't included in the original plans. That's going to change the equipment requirements and the design requirements.

Get your people to be clear up front about what they're planning to do on the network, or two bad things can happen. You tell vendors that you're going to do a whole bunch of stuff that you end up not doing, which means you deploy a lot more network than you need and pay a lot more for it than you had to. Or you will have under-scope the project and don't put in enough network. Then everyone's disappointed because the network doesn't do what they expected. If you get halfway into the process and you don't know what you're getting, it's hard to say what success is.

4. Are there lines that vendors should not cross in their working on a joint venture such as this (many people feel there should be clear church/state boundaries between business interests and the "people's business")?

Williams: Absolutely. There are obviously government officials who are held to tight ethical standards that are more stringent then you find in commercial industry in general. Vendors need to adhere to these standards, respect the purchasing process and don't try to do something that will circumvent it. If you have an order that's going to be for 'x' amount of items and it's above a certain authorization, don't split that in two to try to hide the fact. You have to get the proper approvals.

Vendors have to respect 'Quiet periods' because if you're in there lobbying during this period you'll get thrown out. Or if you get the business you're going to raise a protest from other vendors. That really is sort of the checks and balances in there. Virtually all of these processes have some sort of a protest mechanism so vendors can protest if they think they were treated unfairly. This is the mechanism to ensure that people stay in line.

Tropos had a city customer who had hosted a number of visits from prospective vendors. We wanted to take those guys out to lunch. We could invite them to our facilities and share food brought in for our employees, but we couldn't take them to a restaurant. These are the kind of rules that you have to abide by.

Reinwand: For us it's always been standard practice to do everything by the book, and it's important that cities do this. They should always open the windows to an RFP or and RFI process. Ethical practices should be part of the way every vendor does business. In cities where this is not the case, you're seeing a lot more crackdown on this type of activity. People are more aware of the ways that the rules can be bent, and they are intolerant about this.

I like the city setting strict requirements and then having the option for alternative proposals to be submitted. It's an interesting way to leave the door open by saying that 'we have strict requirements, but we're not opposed to hearing different ideas.' As part of the RFP process everyone has a chance to offer other options. That way, if the city accepts an alternative proposal, competitors can't say later they didn't have the opportunity to do the same thing.

As this technology evolves, there are lots of ways you're going to see of doing these implementations. While cities will be thinking one way today, two or three months down the road as bids start to come in, they may see something in an alternative proposal and say 'gee, I never thought of that. That's a fantastic idea.' You want to let the marketplace have that type of option as long as things are fair for everyone.

5. Can vendors get caught up in the politics of the government structure? How do you minimize the impact?

Khaliq: I believe vendors can minimize the 'politics' of municipal broadband by helping cities shift the business model to focus on the optimization of government resources. We're seeing Dept of Labor statistics that show the number of mobility or non-desk focused jobs within the government sector is going to double the rate of traditional jobs over the next five-to-ten years. This means you need to optimize those employees in terms of what they're doing. You want social service workers out in the field seeing as many clients as possible, not have them focused on paperwork or sitting behind the desk. The same holds true for inspectors. The increasing interest in broadband wireless makes this an opportune time to get good returns on government dollars invested in technology.

Reinwand: For us so far, the way to approach this is to make it very clear what the roles are for the city and for the network provider. When you draw the line clearly, it helps us avoid politics. We don't have to convince the city to do this or that. They understand what their needs are. Many cities get mired down a little in internal politics: the telecom department doesn't get along with parking enforcement or something like that. Once you're operating, be sure to maintain clear expectations for the network. You can minimize concerns of council members if you work to establish clear performance criteria, have clear service level agreements, guarantees on remedies and clear lines of communication.

Philadelphia has been successful because they talked a lot about the benefits. I don't think any council member disagrees with the fact that this is going to help improve education, improve the lifestyle. A lot of their concerns were around how much this is going to cost the city. 'How are we going to make sure the provider does that what they say they're going to do?' We've supported the project champions in the cities by providing them with information that helps address the issues with council members.

Williams: Part of this goes back to respecting the bid process. The bid process is supposed to somewhat remove the chance of political favor. The bigger issue is that someone like a mayor or city council holds the purse strings, so someone needs to champion the project who is close to these individuals. It's similar to a corporate sale where you need to understand where your power base is and do everything you can to shore up your champion. However, a debate within a municipality about installing a network can get a lot of press. This is a substantial difference.

You have to walk a line. There's a set of advocates you have to shore up, there are some people on the fence who you're trying to convince to join your side and there are some people who are dead set against it and always will be. You try to find a way to neutralize this third group without making them a mortal enemy. It pays to be savvy in knowing how to navigate the politics, particularly when there's a big whirlwind caused by the battle of the elephants.

I think that now most people buy into the concept that the technologically will work. They're arguing more about whether it's an appropriate business for the city to be in.

Newman: In a way, you can't get around the politics. But one way for vendors to minimize some of the hassles is to align with politicians who are on term limits because they're motivated to get things done before their terms are up. Democrats, Republicans, they're all the same. They understand that this wireless network brings economic development and it helps them be more competitive. So they'll take up the cause for muni WiFi and be out front with it. The vendors can stay in the background.

Dolmetsch: Yes, vendors can get caught up in this. Normally the politics will be with a localized carrier, or between different agencies within the government that don't agree. We normally mitigate this with contractual statements that say if there are delays due to the city or working with the city, the city will have to bear the costs. We tell them that upfront. We respond to an RFP with certain assumptions and if there are any change orders, the client will pay. It does minimize these political issues dramatically. And we have not had a problem with government administration changes.

6. For vendors, do you expect there to be significant operational differences between delivering products and services to small versus large city governments?

Reinwand: Working with the city and going through RFP process is a time consuming and sales heavy process that requires a significant investment that you're hoping to spread out over a large market at some point. If it takes just as much time to work with a town of 1000 as it does with a city of 1 million, you probably prefer to work with a city that's larger.

You have great benefits in cities that are dense. If you look at a city like Miami Beach, there are a lot of people packed into a small area. Even though it's a small city, it's very attractive because the economics of how many people you can serve per radio goes way up. Currently, we prefer to work with large dense cities, but don't rule out small cities that have similar characteristics.

With communities that are more spread out, you have to start looking at different type of technology. A mesh WiFi network with its typical access points isn't going to make economic sense when you can only cover three or four house per square mile. The operational requirements are significantly different. You have to look at things such as fixed wireless. Provisioning is different. Professional installers have to come to the homes, climb up on the roof and mount the equipment. In order to get that installer there you have to schedule an appointment.

These operational complexities make the business model more challenging. This isn't to say that it can't be done. The economics are just on the verge of making it viable for companies to come in and address these issues.

Williams: As a matter of course, the larger the program, the more complex and more difficult it is to manage and the more carefully it has to be managed. There's a reason why Philly made headlines in a way that Shasta, IN didn't. When you're working with larger cities the visibility is higher, and the media-level politicking is much higher

When looking at vendors, the size of cities in which they have deployed and their expertise in the technology are interrelated in some ways, but scalability is still a big issue. Some vendors' systems are inherently more scaleable, allowing them to roll out to larger cities. When you have technology that allows you to scale to a larger city, then you're more likely to have success. Smaller cities can work with less robust technology, but at their peril.

In the end, the goals of a city are more important than the size of the community. If all they want to do is light up a few blocks in an area downtown to try to attract people into their business district, then this is less technically challenging than rolling out to 10 square miles. There is certainly a greater variety of vendors whose products can work in the few-blocks scenario than in the 10 square miles.

Dolmetsch: No, it's not much different as far as physically putting up the network because even when you do a big city, it's broken up into chunks. You never look at this as one big picture. It really is a quilt, whether each part of the quilt is two square miles or 20 square miles. You have a grand plan, but it's always easier to bite it off in smaller pieces.

In smaller cities, the RFP process is a whole lot less political. For one thing, there are fewer politicians. Also, their budgeting process is much more defined. There is often one, maybe two decision makers. In a bigger city there may be 10.

Don't forget your marketing campaign

Though it may not be the first thing that governments consider when launching a major technology initiative, you need to have a good marketing plan for municipal wireless. The reason to do this is political. In Chapter 5, I discussed the need of building support among your various constituencies that you cultivate into a political lobbying force. The marketing campaign is a method by which you build that support, and this should tie into your focus groups, pilot launch events and other consensus-building activities.

Personal marketing efforts can help with those elected representatives who are supporters or are still on the fence as you move from pilot to RFP process to full deployment (die-hard opponents rarely change, so try to neutralize their influence). Richard Miller of Innovation Philadelphia advises the approach of making it easy for representatives to embrace the initiative as their own and look good for their constituents.

Mayor Sangiovanni of St. Cloud, FL included an education campaign in that city's marketing efforts that brought along the public as well as council members who weren't directly involved with the initiative as he was. "It was they who had to make the big decisions about putting several million dollars into this capital expenditure. We broadcasted meetings on our government TV channel, ran ads in the paper, conducted workshops and so forth. The political side was brought along with the discussion at the same speed as the various constituents. And we didn't spend a whole lot of money on the entire campaign."

Be consistent in your message. Every marketing effort since the beginning of time that has achieved great things was driven by relying on an easy-to-grasp message that rallied people to buy a product or to change a government or a government's way of thinking. T-shirts during one of the pilot launch parties stated "Get unwired. Stay connected. Reach the World. WiFi is Philly." Be persistent in getting your message in front of people. All the means of mass media, from the Web to wall signs are at your disposal. Use them. Speak the language of the people, avoid tech-talk and geek-speak.

Always end your message with a call to action. When you finish your pilot projects and the evidence supports the value of the technology, channel your constituents to take action in all those ways that make sense relative to your area. I read a blog relating to taking civic action where the writer said "This legislature is the beast we created by our apathy." Though not a reference to the WiFi debates, it nevertheless points to how it is that good ideas are often sacrificed to bad laws. If you are truly committed to the value that WiFi technology can bring to your community, then you have to rally your community to take political action when necessary.

Moving forward with mobile workforce applications

When it comes to the mobile workforce applications, the transition from pilots to full deployment is a lot less political than deploying the broadband network. Your staff and workers are all employees who you have some amount of control over. However, don't become complacent that this control factor is going to solve all implementation challenges. You still need to exercise expectation management, use a little diplomacy and practice some internal politics. In organizations across all industries and types of government, a common denominator with deployments is that younger employees embrace wireless and its resulting changes easier than older employees, and 20-year veterans nearing retirement can care less about your vision or technology.

Gene Smith, Business Program Manager for Xcel Energy, an oil and gas utility in the mid-west, was involved with a mobile maintenance and general work management for 1300 field technicians. Their approach to the pilot-to-deployment transition offers some good lessons. For the pilot projects that were targeted for production, they had two teams working together. One team was responsible for successful execution of the pilot. The second team worked with the pilot team and was responsible for identifying activities and resources that would be required to take the pilot to a production-ready application. This reduces the time and hassles moving between pilot and deployment of the solution.

To ensure that expectations are well managed, they presented the program for their wireless initiatives to the highest executive levels for approval. Within the program were specific projects. Each project was required to have a plan which includes ROI analysis, request for money and resources, vendor requirements, scheduled timelines and so forth. Everyone was clear on roles and responsibilities.

"When we begin execution," Gene states, "everyone involved will use Mercury Interactive project management software which maintains and distributes job summaries, status reports, milestones, costs-to-date and practically every other detail of the project. If you go over budget, don't meet objectives, or otherwise stray from the plan, then the software flags this and makes sure people get back on track. It's a rigid program that keeps everyone in line."

Here are a few tips and guidelines for keeping this part of your deployment on track.

Can you really get there from here?

Give or take a few days, one woman delivers a baby in nine months. But you can't get nine women together to deliver a baby in one month. However, when it comes to technology, some managers move from pilot to deployment with the nine women/one-month philosophy. "If we generate $10,000 in time savings from 50 employees in a pilot, it's a given we'll generate $100,000 when we deploy to 500 workers."

Well, maybe you will. Maybe you won't. This is not to say that you won't generate enough benefits to justify the investment. But if senior directors spend a year or more expecting $1 million in ROI and you deliver $400,000, there could be some long faces around the table at performance review time.

Unless all mobile workers have the same skill level and do the same job in the same amount of time, the revenue increases, time savings and productivity gains may not be directly proportional to the increase in users from pilot to deployment. And it's likely the costs for everything from the logistics of deployment to the overhead of managing users won't increase or decrease in direct proportion to the increase in users.

This is an important point because the time between the pilot and moving to full deployment is your last best chance to make changes in projections about how much you're going to gain with this application and at what price. Once that deployment train leaves the station, it is much more expensive to respond to any major miscalculations in pre-pilot assumptions.

How much time should you spend on this analysis? It depends. How closely will you be held accountable for the projected ROI or job performance numbers? Some organizations are happy if they just see any marked improvement in the metrics which are used to determine success, while others want to see specific numbers met after full deployment. The latter case is where you'll have to spend lots of hours with calculators and spreadsheets.

A greater interest may be in whether or not you can complete the deployment within budget. What happens when each cost is extrapolated out to 500, 1000 or 3000 users? What associated costs come into play? It may not be expensive to provide replacement parts to 50 users selected from one department for a 90-day pilot. But 2000 workers in large geographical areas can drive the annual logistics costs for replacing parts and repairing devices through the roof.

Just as important as it is to accurately extrapolate overhead costs and ROI, you have to do the same thing for problems, particularly technology problems. During a pilot the tendency for some is to say "that's a minor problem. We can deal with it later." They don't want to delay the transition or they dread having to do another pilot. But as the commercial for a chain of auto repair shops said, "You can pay me now – or you can pay me later."

If you encounter what some consider a small glitch before deployment starts, look at how much time it will cost for people in the field to work around the problem. One or two thousand people losing a couple of minutes each day becomes big bucks at the end of the year. Or what you consider minor might be major to the person in the field or remote office. Estimate the cost of bringing your mobile workforce's devices back from the field to fix the flaw. Pay now, or pay a lot more later.

Managing people effectivelyis crucial

1. Know everyone's limitation.

Wireless is hip, happening and sexy, so a lot of people want to be in on the action. But on the business side and the IT side, you have to resist the personal or political pressure to put people in positions of responsibility on the project team unless the positions match their capabilitites and availability. Likewise, you have to keep higher ups from directly controlling deployment processes for which they have no expertise.

IT people overworked with projects don't need day-to-day programming and network testing responsibilities for a new wireless application added to their plate. Rely on the vendor or systems integrator for this. Business managers who haven't been on a field service assignment in 10 years shouldn't be writing the technical requirements to give to IT – at least not alone. Appoint someone to guide your deployment who has the ability, resolve and authority to make the right people assignments, to remove or re-assign key people if they're not cutting it and to take projects outside of the organization if necessary.

2. Put someone in charge who knows the technology – or is willing to learn.

Relatively few managers, executives and even IT people are experts in mobile and wireless. They don't necessarily need to be. But they do need to develop a working knowledge of the technology, be aggressive about asking questions of IT pros, vendors, consultants, etc. and not have ego issues with saying "I don't know."

3. Don't undervalue end users' feedback.

One of the skills executives and project managers need to sharpen is their ability to effectively solicit and properly use end user feedback. Several people have commented on how they get maximum user acceptance of the application by responding to their feedback early in the pilot process. And if failing to do this, at least get that feedback the minute you sense a problem is brewing with your deployment. The good manager knows that often the solution is not on their boss' desk, but in their workers' hands.

4. Minimize the role of those who manage too much or too little.

Once you kick the deployment into motion, particularly one that involves many hundreds users, there are people and materiel logistics involved to rival the Normandy D-Day invasion. This is the time to scoot the micro managers to some other part of the organization. William Tara, CIO of American Medical Response (AMR) hires professional project managers on staff and lets them have at it. Greg Lush of the Linc Group lets his people do their thing, but he will step in if the project starts to miss deadlines or stray from the plan. You have to find that middle road between too much and too little management that best fits your situation, preferably before the project starts than after.

Managing expectations – is a never-ending job

The demon/angel that hovers in the wings of every deployment is Expectation. Civitium's Greg Richardson says "Frustration is a function of expectation." When reality meets Expectations – of deadlines, budgets, ROI, product benefits, productivity, technology limitations, etc – you either have bliss or hell on wheels. Here are a couple tips to help you experience more of the former than the latter.

1. Whatever you think, double it or cut it by a third.

When dealing with relatively young technology such as wireless, assume that everything takes twice as long and costs twice as much, and often the results expected from the technology will be less than what you eventually get. There can be as many reasons for this as there are organizations. But nevertheless, bad things happen during deployment when someone(s) important who expects or needs the project to cost "x" or be finished by "y" and produce "z" and is disappointed with what they really get.

Remember, no company ever went out of business and few people have ever been fired for a project that finished ahead of schedule and under budget. The Chicago Mercantile Exchange's first wireless application was a costly disaster. But when their project group launched a second wireless application a couple of years later, they played down expectations and the project was such a raging success that they couldn't hand out PDAs fast enough.

2. Skepticism in appropriate doses is good for the bottom line.

Sales people have one primary mission, which is to sell things, though keeping the customer satisfied is hopefully a close second. And more times than the technology industry likes to admit, some sales people don't know a lot about the technology they're selling. So, until you are thoroughly convinced that the sales reps sitting before you have your best interests at heart and a better than passing technical knowledge of their product, be relentless in determining if their products are likely to perform as planned. Tom Kilcourse of the Ratner Group, after being burned by a previous vendor, went to the next vendor's offices, met with their top executives and took his IT director along with him.

Granted, the pilot is supposed to uncover potential problems, but there are so many different technology dynamics at play in large deployments that you never know how an application is going to scale until you get in the middle of one. So do yourself a favor and insist that vendors have contingency plans for possible problems as well as guarantee they will resolve any problems. Being the skeptic, don't assume they plan to bear this cost. Ask. With multi-vendor projects, expect finger pointing if any major breakdowns happen. If you can, develop some service agreement language that protects you in this scenario.

3. Be honest about what vendors expect from your organization.

Often vendors' sales people are not at fault when organizations end up with applications that don't perform as advertised. When an organization doesn't clearly articulate what they want or expect, or the pilot doesn't adequately uncover what their mobile workers expect, then the resulting application often gets poor results. If the vendors expect, based on what you tell them, that you have perfectly running business operations, that's what they'll build to. However, if your operating procedures are in dire need of improving, then the application you get probably isn't going to produce the results you need.

It has happened that some organizations don't like to talk in too much detail about their inner business workings for security reasons, which is understandable. But you should make a decision about how you're going to deal with the potential of getting results that are less than what you want because you don't want to give vendors the details they require to build what you need.

4. Be prepared for Murphy's Law of Unintended Consequences.

Murphy's Law says that what can go wrong will go wrong. His Law of Untended Consequences of Technology says sometimes things go so right that other things go wrong. One group was successful in reaching their application's goal of reducing the need for delivery drivers, but they quickly started having an unexpected loyalty and retention challenge with the drivers who were left. Another company increased the efficiency so much for their field workers that shortcomings in the home office operations suddenly had to be addressed.

You need to plan for change before you start the pilot process. When an application does what it's supposed to do, you can be in the middle of a deployment and be surprised when the floodgates to increased data flow, greater workloads for office staff, ideas for new applications and so forth burst open. People expect some success, but without proper planning it arrives at higher levels than they expect. You may need to slow down the forward momentum of the deployment so the department can adjust to this, or IT may need to modify the application. These changes can be costly when you're not prepared.

Take advantage of early successes

Dealing with the occasional trials and tribulations of internal politics can be a cross for the project team to bear. To make this challenge more bearable, it pays to take advantage of – and heavily promote – early successes. This gives you a little breathing room and maintains executive-level support while waiting for the deployment to finish AND to see some of the benefits. Depending on the number of office and field staff involved, it can take a year or so to finish the deployments. Some benefits might be immediately apparent, while it can be many months for other benefits to be quantifiable.

1. Know what the business manager's hot button issues are – promote them first.

You may be ecstatic that the servers allow 500 mobile users at a time to use the new VPN system. The department director's first concern, however, is that the old system required eight steps to get through the security procedures and this killed time that workers could have been inspecting construction sites. Is the new system going to eliminate most of these steps?

2. Don't underestimate the little things.

You IT person may see the big picture of the technology's potential impact. But to a harried manager, the ability to get two paragraphs of data delivered to a PDA five minutes before a crucial meeting with the mayor is the be-all and end-all of their world. In the early days of deployment, promote the technologically trivial things that affect higher ups personally in terms of their ability to do their own jobs. These may be easy successes to replicate throughout the rest of the department.

3. Everything is a potential early success to be promoted.

Hardware distributed on time, workers trained ahead of schedule, an increase in data that's captured on network servers. Success at every level is positive internal PR for the project. Just don't overdo it. A weekly update in the first couple of months of deployment listing two or three simple items that the average non-techie can understand should be fine.

4. Paint a picture of great glories.

After accumulating a series of initial successes, show senior administrators how these initial successes are the foundation for greater benefits that will make yours a leader among municipal or county governments. Being a leader stokes the city's – and officials' – ego. This ego equity becomes valuable political equity when you're eight months into a year-long deployment, some of the greater payoffs are still a ways down the line, and a new administration might replace some of the directors who championed the project. But if you're producing clear wins early and can tap into visionaries long-term thinking, you'll do well getting over the middle ground.

5. Create an early warning system.

As eagerly as you want to claim bragging rights for early successes, you also want to nip end-user resistance to the technology in the bud. The pilot process should uncover and address most of the issues that can cause this, yet somebody somewhere will get a burr under their saddle about something. During the first stage of deployment have a feedback mechanism of some sort linking users directly to IT so you uncover, acknowledge and quickly respond to any complaints. Don't let one vocal individual incite a chain reaction.

To sum it up

Once you determine it's time to transition into full deployment mode, usually after the RFP is awarded, be ready to move with conviction. It helps a lot if you start planning the steps for deploying mobile applications while you're in the middle of pilot tests, and as soon as makes sense for the network infrastructure. Allow for some level of flexibility.

As you start putting the technology in place, don't forget to manage expectations well, and execute a smart, aggressive marketing campaign to all of your constituents and stakeholders. You have a long year ahead and you don't want any disappointed folks at the end of the rainbow.

Chapter 11

TCO – the stealth threat to a successful deployment

In discussions about mobile and wireless application deployments, you hear a lot about ROI – return on investment. Why it's important, how to calculate it, why it's difficult for governments to measure it since to much of what a government does is not about generating revenues. For municipalities such as Philadelphia that are pursuing muni WiFi to achieve economic development and social goals, determining ROI is definitely not a clear cut process from the perspective of financial analysts.

What often escapes attention in the discussion about ROI within all industries, not just government, is TCO – total cost of ownership. TCO is everything that it costs you to buy, deploy and manage the application during the life of the application, which is more than the price of hardware, software and upfront services. TCO directly affects the value that you derive from a deployment and can be a nasty surprise if you don't make plans to keep it under control.

One reason TCO catches people unaware is that there aren't a lot of wireless deployments fully completed, and wireless is such a different beast to implement compared to other technologies. This chapter tackles some of the main contributors to TCO, starting with how it affects municipal WiFi infrastructure deployments and then looking at the TCO for mobile workforce applications.

TCO is not a deal killer that forces you to put wireless applications on the back burner, just a series of line items that need to be factored into the budget and budget management. Assessing TCO requires extensive feedback gathering to find out what's needed to make the technology fully operational for and by your end users – training, accessories, upgrades, replacement parts and so on. You also need to thoroughly research where technology capabilities and feature sets are moving.

221

There's no formula for calculating TCO. But there are some main issues to consider and questions to ask within the organization and of the vendors you use. Don't let this discussion slow you down, just become wiser for it. It will help you anticipate, adapt and respond

Understanding total cost of ownership (TCO)

When you buy or develop typical software applications these days, most of the hardware is already in place. You pay the purchase price, an added amount for training and services, and maybe factor in some operational downtime during the installation. Wireless applications, on the other hand, are a world unto their own.

When deploying a citywide WiFi infrastructure there may be unexpected costs that add to your TCO, such as changes in the physical landscape, difficulty acquiring rooftop access, or the bankruptcy of one of the vendors. Underestimating the time required for permit approvals can result in build out delays.

With mobile workforce applications, devices are different from one another in operating system, size, capabilities. Subsequently the cost to deploy, manage and distribute them is higher than if giving everyone a generic PC. Hardware replacement costs are higher because smaller mobile devices such as PDAs and tablet PCs are dropped, lost, stolen or wear out faster than desktop computers. There are accessories such as cradles, cables, bar code readers and WiFi cards that are easy to break or lose, so replacement costs for parts for a large workforce can stun you. Security is a huge concern for both the network and mobile devices, so the price of buying and managing security products adds to your TCO.

Since most mobile workforce applications are comprised of components from different vendors, it's hard to see the whole picture in terms of costs if you don't have a good systems integrator. Training can take longer and be more expensive with mobile workers. They're away from the office for most of the day, so when they have a problem or forget how to do something, they can't walk over to the cube of the office techie which is what office staff do.

All of these factors impact what a deployment ultimately costs. Existing back office software may need to be modified to accommodate the mobile software. A big TCO inflator for both the network infrastructure and mobile applications is obsolescence if you don't plan properly. Most of wireless technology is in a continual state of evolution.

TCO is sometimes the cold water thrown on the party after an application has been installed amid promises of reduced costs and improved productivity, then evaluated later to validate those promised benefits. When organizations are unprepared, TCO is the torture of a thousand paper cuts as little surprise costs pop up from the very beginning of an implementation. All of this directly impacts your ROI since ROI equals the value of the benefits derived minus the costs of putting the application in place. TCO also can come with a high political price if you don't tell people to expect these kinds of costs.

The good news is that many of these expenses can be contained or eliminated if you become aware of them during the pilot. If not contained, at least you can budget effectively for these costs so your ROI projections are more realistic.

The price of "Free"

Brad Tabaac, the owner of Friendly Pharmacy in Philadelphia who participated in one of the focus, was right when he said "nothing's free. It's has to be paid for by somebody somewhere." Earthlink and Google are making headlines as they offer to build the WiFi network at no cost to cities. When you look at the TCO for deployment of the infrastructure, here are some things to consider when you get that offer for a free deployment.

In Philadelphia, the infrastructure build out is free, but everyone including the city will pay for access. In other cities, service providers are offering to do the build out and provide access to city employees for free, and if Google does San Francisco, everyone gets it for free. But how much is the access service if it's sold to constituents, and how does that pricing fit affect your ability to execute economic development or digital divide projects? This needs to be communicated to the community. Also, what happens if the service provider one day decides that the effort isn't profitable enough for them, or what if the question "who really controls the network" turns into an expensive tug of war between provider and city?

Think through all of these contingencies thoroughly before launching deployment. Examine the business model of the providers making the offer. How are they going to make money, and how will this impact service? Craig Newman of Motorola cautions that a free network for all of your citizens sounds really good. But when you offer a free network, what's the incentive to upgrade? You just throw it up there and whatever happens, that's good enough?

"The other issue is who ends up manning the call centers to help users needing assistance? If the provider business model relies on revenues from ads, which is an un-proven source, they might have a small group assigned to customer service. So then customers get upset because they're not getting good service and call the city or the PUC. All of a sudden the city is carrying this cost."

Even if you resolve these peripheral issues around of free deployment and free access for citizens, there are still costs. "If cities plan to provide access to make government mobile workers more efficient, they have to take into account the hardware," states John Dolmetsch of Business Information Group. "We did a deployment in Queen Anne's County, MD where the cost of the equipment to go into vehicles was more than the cost of the infrastructure by 200%. So it's one thing to promise the world, but people need to look at the true costs if they plan to do anything with the technology."

This holds true for economic development and digital divide programs. There are federal grants and corporate community service programs that a municipality can tap into. But even so, without the extensive creativity and hours of sweat equity (with attendant payroll costs) a free network alone pays little dividends in social change. Wireless Philadelphia is structuring its deal with Earthlink so the non-profit shares revenues generated in the city by the service provider to offset some of those people costs. You may want to consider a similar arrangement.

End-users may carry some costs as well with free access. As Greg Richardson observes, the digital divide charter influences WiFi technology on the network edge due to the cost of the customer premise equipment (CPE) to boost the radio signals. "It costs $50 - $75 for the bridge you find at an electronics store. In areas where there isn't the density of pole fixtures to hang radios, you have to look at more expensive CPE options such as the Canopy subscriber unit which is $400 or $500, plus a technician installation charge. With an average monthly revenue of $10 per low income user, it would take you 60 months to recover the investment. Or you have to say to a low income person 'I'll sign you up or $10 per month service, but I'm going to charge you a $500 activation fee.'

There are some other operational TCO figures that you need to consider whether you are directly paying for the network deployment or not. "The communities in the larger cities need to factor vandalism into their budgets," Dolmetsch adds. "Most of them don't. Just as with surveillance cameras, people shoot the transmitters, or climb up the poles to unplug them. In Philly we've been down there three or four times for radios that were unplugged or moved around. It's a cost to go up that pole again. WiFi is fairly new for cities, so it may become less of a problem in time. We see a lot more of this with the surveillance cameras in the downtown areas of some cities. We've actually had to put up bullet proof shields on a few cameras."

He also warns about the cost of maintenance. Cities don't ask for 24/7 maintenance on the network. They expect this to be similar to phone systems where if something breaks the phone companies come out right away and fix it. Sure, companies will come out and fix something if it breaks, but that's going to cost you extra unless you have a fixed fee contract up front.

The private cost of free

One of the big issues that you need to address is the perceived and actual cost to individual privacy as the result of a free network. Resolve this in the contract, if not before, and be forever vigilant about it as you move forward.

What happens to personal data that is collected by a service provider when people sign on or as they use the network? Where does it go, who has access to it? It's no secret that a lot of people are suspicious of governments' involvement with communication that touches them in a personal way, and theoretically can follow citizens wherever they carry a mobile device. They'll be more wary if they think a service provider is offering free access so it can get free reign to customer data. You don't want to risk the good works that the technology can enable by not demanding clear privacy guidelines about citizens' and businesses' data.

Another factor is the thought that a company such as Google, which makes the bulk of its bucks through selling ad space on its search engine, can use the capabilities of WiFi to target ads to people based on where they are when using the network. As an average citizen, it's one thing for me to deal with ads if I choose to use their search engine or if I have a choice to subscribe to muni WiFi service with or without ads. But if the only WiFi option is one with mandatory ads AND location-based marketing, this is not going to be well received by some folks.

I made the suggestion in Chapter 3 that a sports team or any company with the type of mobile marketing vision of the Carolina Hurricanes could underwrite the cost of the infrastructure as part of its branding campaign and ability to run opt-in promotions. If a city looks at funding this WiFi initiative with creativity there are many business funding options the city has at its disposal that can benefit the companies, but minimize the potential for intrusive marketing.

Regardless of what type of business arrangement you strike for the network build-out, security of the individual and their personal data must not be compromised in the bargain.

Avoiding the TCO impact of obsolescence

When discussing WiFi networks, a frequent TCO topic of conversation is technology obsolescence. If governments don't plan properly for the frequent fadeout of one technology and the introduction of a new, improved technology, there can be financial hell to pay. Some aspects of WiFi technology have become stable. For example, while different flavors of WiFi unfold, it's almost certain that every laptop comes with a built in WiFi card and these cards are dirt cheap for desktops and older laptops that don't have them. Many of the new models of PDAs, smartphones, tablet PCs and even some cell phones are coming off the assembly line with WiFi built in or WiFi cards readily available.

As for the network itself, many elements of the technology are still evolving, a couple have settled and you never know what's going to pop up in the next 12 months to either reinforce standards or create new ones. From this perspective, $15 - $20 million is a lot of money if you make the wrong choices and have to re-do or replace large parts of the deployment. Let's look at what some of the experts have to say about how you can avoid or significantly minimize the TCO impact of obsolescence.

Proper previous planning is key

"Put your money on the technology that's in the digital device," advises Intel's Paul Butcher. "Bet on technology that's walking around with the greatest number of people. Build infrastructure behind that technology. It's a good bet that WiFi will stay in future computing devices, as well as be introduced into many handsets, and WiMAX will start to be integrated into these devices. Intel will drive down the cost of WiMAX technology with our fab that goes into these devices until it becomes a $5 part that's in the devices when people turn them on."

Dianah's team put a requirement into the RFP that the winning company has to make upgrades as the technologies changes. Both you and whatever group manages the infrastructure must be prepared to address the issue in some way. This means 1) the service provider and vendors underwrite the upgrade effort, 2) you have a sustainable business model with revenue streams, or 3) the network generates strong subsequent cost savings in government operations that enables you to fund upgrades.

Besides the potential obsolescence of the technology or technology standards, there is also the issue of individual vendors' products becoming out of date. As I mentioned before, IT typically only has to deal with one vendor, maybe two, in the average technology deployment. Things are clear as to what you can expect. With wireless there are multiple vendors and ISPs and their respective pieces of the technology puzzle evolve at different speeds.

"This is one reason why you have a prime contractor or system integrator," remarks Bert Williams from Tropos Networks. "There needs to be someone on an ongoing basis who will advise you what new technology is available and how it can be successfully integrated into the network."

One view that I've subscribed to for a long time is that you don't need to change technology just because it's old. I toured the plant of Itronix, a rugged laptop maker, and they had some machines for Sears in the service repair area that were pretty old. But since the laptops kept doing the same job year after year, Sears wasn't interested in swapping out the old units. Bert adds "as long as a technology continues to support the apps that you want to use, it's not obsolete. There shouldn't be a rush to adopt the latest tech just because it's available. Evaluate if the network that's running today is doing what's required.

Also consider that, even though technology changes rapidly, people over esti- mate the pace at which new technology will be widely adopted. One criticism you might hear from people opposed to the network is WiFi will become obsolete in a year when WiMAX becomes available. Their implied or stated argument is that here's another reason cities shouldn't deploy any kind of network now. "Let me tell you, there's something like 160 million WiFi clients in the world today," counters Bert. "There's virtually zero WiMAX clients. Those 160 million aren't going any- where anytime soon. Serve the base that's out there now rather than what might be adopted in the future."

When you're evaluating vendors take into account whether they have a group of staff monitoring these technology changes. If so, they're familiar with navigating past the dangers of obsolescence. On your end, you have to understand applications and the requirements of the network. Have reasonable expectations about the ability of the technology to run those applications and what it's going to cost you to get there. As long as all of this is explicit up front with everyone involved, then you have the basis for a sound implementation. But if you don't set the right expectations, you're going to have problems the whole way through.

A question that might come to mind is, what do you do if you get into the middle of an implementation, or maybe a couple of years down the road, and realize you haven't budgeted properly for obsolescence? John Dolmetsch of Business Information Group believes that if a city does this wrong, they can miss out on opportunities to create a greater and shorter ROI model. It's not just about WiFi. They can be too shortsighted and not plan for a five-to-10-year infrastructure evolution and so the network won't be able to scale effectively. They will have sized the network too small and won't be able to support additional services on the network. This results in spending a lot of money up front and you still have to do an expensive upgrade later.

If you write an RFP just to support WiFi, the network may not support higher-end connectivity services and other future needs of the entire infrastructure. There has to be a lot of forward thinking on the part of the city to support some of the things in your future such as RFID, location based systems, the ability to wirelessly track missing kids, and other societal benefits. How do you build the network today so that it's ready if someone has to come in and add additional backhaul capacity, a different frequency or newer technology that's going to provide a certain type of infrastructure?

It is the vendors' role, especially in it's response to an RFP, to offer a flexible solution and point out the short-comings in an RFP. John concludes that "if they don't see certain things in an RFP, vendors should say 'you really need to be thinking about this. This is what it's really going to cost to add this kind of support.' You don't want to go in there and just throw up a barebones system. Typically when we respond to RFPs and we find items aren't included in there that should be, we throw them in as options along with explanations about why what we're proposing is a better solution. It's not a hard thing to withdraw the RFP after it's been issued."

Avoiding obsolescence with workforce applications

You have to plan how to avoid obsolescence with your applications as well. The shelf life of many mobile devices is a little less than a year before being pushed out by newer models. Smartphones are recycling in and out of popularity within months. Ruggedized devices have a longer life.

Integrating new models of devices can be expensive if a pilot runs for more than six months and then full deployments run even longer. The model you test may not be available as full deployment begins, so you need to do additional testing to make sure the software will support the newer model. This extends the time of your pilot and adds to the costs, though better this than actually deploying the newer models and then discovering that they have problems running your application.

One way to minimize the obsolescence factor is to design or configure software applications during the pilot so they don't require the hardware to have "bells and whistles" features, or even the same operating system. You could get years of use from the initial hardware purchased because the applications require just the basic computing capabilities of the devices. This hardware independence in the software also allows you to add newer hardware models or devices from different vendors to the same application.

The middleware product Philadelphia uses allows L & I to concentrate on delivering integrated services while requiring relatively little new software coding to deal with back-end changes. The services that inspectors, other departments or customers access are seamless. Now they no longer rely on particular data sources that used to be updated only weekly, but get instantaneous updates as transactions occur. A good middleware product enables you to keep the front-end user interface the same and all you have to do is change the back end. If another product comes along that's better, they can add it easily without creating a new database.

That said, L & I also does a good job of staying ahead of the hardware evolution curve. "We've gone through three production servers since we initiated this project a couple of years ago," Jim Weiss reports. "We monitor transaction volume and try to stay six months ahead of demand. Our handhelds have a quicker turnover than the annual change in some desktop computers." People in the field put more physical stress on the hardware going in and out of locations where the device gets banged around, constantly going from high temperature to low temperature environments, high humidity to low humidity. People in other municipalities are pushing the hardware market as much as Philly does.

Commissioner Solvibile adds, "Since we first started we went through three or four handheld models. When we go live in February, I'm sure there'll be something different out there. We haven't bought the equipment yet for this last group of users. This is on purpose. We're holding off until the last possible moment so we get the most up-to-date, efficient piece of equipment."

Paul Moen is the Software Designer-Developer for the North Dakota State Water Commission. He managed the deployment of five applications to enable mobile data collection and reporting by state-managed aircraft pilots so the agency can better study the effects and impacts associated with cloud seeding. From his perspective, "the technology is sure to change as your development continues. Try to keep up with it, but only to the point where it is detrimental to you if you don't. This all depends on many things, such as the software you use and how easy upgrades are. Never upgrade or use new technology in the field until it has been thoroughly tested in the lab."

Technology should be used to improve your application, not interfere with it. The saying "Don't fix what isn't broken" rings true in most instances. Sometimes, there are technology advances that are well worth the time and effort. The most important thing is to never loose sight of your goal and don't let technology drive your mission.

Bring a broader vision to department thinking

Even though department managers aren't directly responsible for the citywide infrastructure initiative, it's in their best interest to have input in those decisions so department applications aren't rendered obsolete by the network. Rizwan Khaliq from IBM advises, "when developing the network infrastructure, look at different types of networks, WiMAX and broadband over powerline as well as WiFi. You'll likely find the technology is a hybrid that's IP based. This allows you to leverage investments you've already made in traditional Internet and intranet technology." Give the city's project team the specifics of your department's IP-related technology that's in place.

A number of customers look to the device first when they evaluate mobile workforce applications. As you dig more into what people want, you find they haven't thought through what they need to do with wireless. Middleware is the key, but look for open source applications so you're not pigeonholed into one vendor's products. That, rather than customizing everything, is the only way to leverage the community of developers and also guard against obsolescence.

Rizwan adds, "your thoughts on mobile devices should go beyond handhelds. 'How else can I use the edge of the network? What's going to touch that environment?'" If your workers are responsible for the management, tracking, monitoring and/or maintenance of numerous physical assets, mobile or otherwise, this network can play a vital role wireless-enabling these assets. But you have to select the enabling technology with careful forethought because this is another piece of the puzzle that can become obsolete."

It pays to look at organizations outside as well as inside of the government sector to see how they deal with obsolescence issues. Tom Kilcourse is the Director of Facilities Management for The Ratner Group, which owns the national chain of Hair Cuttery salons. He has his VP of IT conduct vigorous research to determine if a vendor's software is likely to be around for awhile. If you feel you need expertise beyond your IT staff, you can also bring in a consultant or an industry research firm such as Forrester or the Yankee Group to evaluate potential vendors. The main thing you want to know is what steps your potential vendors are taking to keep up with technology changes.

A lot comes down to there being a partnership between you and vendors. If you have changes coming online with a department's computer network or the city's WiFi network, how does this impact your vendor's product now or down the road? Can your IT people support these changes and also ensure that the vendor can support the changes as well?

"Our field workers don't care too much about having the latest gadgets," notes Kilcourse. "They care more about fast connection speeds that don't drop frequently. We are looking at moving to tablet PCs since it's easy for people to take it with them into stores and prevent device theft from the trucks. On the other hand, there has to be a point where your equipment is kept up to date. The oldest laptops we have are two years old. A lot of organizations have leasing arrangements for equipment so they can continually upgrade devices without going bankrupt. Of course, you then have to deal with the TCO of support costs every time you change devices."

Another situation that some departments may face is the use of WiFi networks within the offices that are in place as the city's network comes online. You don't want these investments jeopardized either. There could be some security challenges or potential hiccoughs with workers' devices switching between networks as those workers move between indoor and outdoor work areas.

TCO management guidelines

To put things in perspective, consider that the cost of a mobile device "represents just 25% of the TCO for a mobile implementation," states Kristi Urich, Director, Field Service Industry Marketing for Intermec Technologies, a rugged device manufacturer. She explains that much of the TCO is for support, sustaining devices' operations in the field, and managing spare units. There's also end user downtime in the field while they manipulate data, or try to troubleshoot and fix problems. If you use this as your starting point, you can create an initial deployment and application budget, then follow some of the following guidelines to identify and possible curtail specific costs.

Let's start by looking at the most obvious contributors to TCO. These are the usual suspects, and any vendor worth doing business with should help you calculate their costs. Some vendors may already factor these costs into their pricing plans.

There are learning costs regardless of how easy a vendor tells you their application is to use. Even if end users can pick up and run a mobile app within five minutes, the people installing and supporting the backend of any major software deployment will need to be trained on these products. Besides costs for instructors, manuals, time in classes and other formal training expenses, people need time to learn through repetitive use how everything works. People have on-going questions, so whether they're on hold waiting for vendors' customer service or calling your IT staff, this is lost productivity time.

Your pilot project should include calculations for learning costs. Once you figure out what these are likely to be, determine if the vendor or your IT people can create intranet content that's easy to navigate and facilitates quick learning/review. Luckily software vendors appear to be making things simpler, but the more features vendors cram into mobile devices, the greater the learning curve.

Software programming for in-house apps, customizing "off-the-shelf" apps, upgrades (both the price and the effort to physically upgrade devices) and end user support contribute to TCO. There are also costs for new security software, upgrading existing security programs to support mobile access, and managing passwords and access privileges on mobile devices for current, future and departing employees.

As the mobile industry matures, there are an increasing number of software tools to help with these tasks. AppForge is a veteran in area of software development tools that facilitate both the creation and deployment of mobility programs. in Atlanta is a veteran in this area. Intellisync (www.intellisync.com) is a vendor marketing middleware and applications that facilitate remote software deployment, management of multiple vendors' devices and security for both mobile devices and your network. The pilot project should assess how well these types of tools control costs.

Tom Baumgartner of the Pinellas County Sheriff's Office describes steps they used to contain costs for managing their departmental WiFi network. "On the IT side, we implemented a centralized managing system that monitors and updates access points at any given time. IT could resolve problems quickly and efficiently by checking alerts on the managing system. We also created a check-and-monitoring feature so we can alert managers to where there is a problem, or if their people aren't adhering to policy. The best solution is some type of 'security management' application, meaning software that controls access and security, such as the Wavelink application we use."

Your application should allow you to regularly check to see all of the access points and where they are physically, detect what has failed, try to re-start failed devices, and so on. Another way to keep system management costs down is to have this software provide sufficient documentation of physical installations of your network infrastructure by remotely going out and collecting site-surveys. This is a basic starting point for security that is often overlooked. As you add new device users and access points, you have to create a 'paper trail' that is part of the oversight process. Always look to add on to this documentation.

The sometimes forgotten TCO factors

In the hustle and bustle prior to deployment, a lot of things can get overlooked or just don't come to mind. If the pilot group wasn't very large relative to the total mobile workforce, or the duration wasn't long, some of the potential TCO factors won't come to light.

Change management. New technology that significantly alters the way people work demands that you adequately prepare to deal with the resulting changes or face possible operational struggles that eat up time, cause employee resistance and otherwise negate some or all of the anticipated benefits of mobility. For example, shifting more data access to the field can result in more decision-making by people not accustomed to this role. If you don't spend time preparing them for this new responsibility, they could make bad decisions that are more costly.

Deploying mobile devices that offer the benefit of greater workforce account-ability for management also could discourage employees from using the technology because they fear intrusion and micro-management. Your TCO then includes the cost of un-met productivity gains. Effective mobile applications for field workers can dramatically increase the amount of free time that home office staff have because driver dispatch or forms processing tasks are eliminated. If you don't plan ahead of time for reassigning them to other beneficial tasks, you risk the cost of having people lounging around or otherwise underutilized.

Back-up hardware. Mobile devices are subject to damage, theft, irreparable technical glitches and eventual obsolescence, plus the administrative overhead and loss of productivity while replacing the units. The longer it takes to physically replace a unit, the greater the productivity loss. The larger and more far flung your mobile force, the greater the administrative and logistical overhead.

Though it may take six to 12 months before you can gage what your replacement rate will be, you should probably budget for 5% to 10% more devices than the number of people you plan to equip. Configure these devices so they're ready for same day delivery once someone reports a problem. This upfront expense should save a lot of downstream costs.

Make sure you take similar steps with vendors for the municipal network infra-structure products. An access point that fails in a densely packed urban area with multiple access points per block probably won't be noticeable. But networks in suburban and rural areas have a greater reliance on each piece of equipment. If your area is subject to major natural disasters, or is considered a likely target for terrorists, keep a back up equipment supply in quantities that can replace the loss of 15% - 25% of your network.

Hardware standardization. Even if you equip employees with the same devices initially, Intermec's Ulrich believes that having mixed devices is inevitable. "The market life of mobile devices is 11 months," she observes. "So by the time you start replacing units three or four years later, the new devices will be different than the originals." Enforcing organization-wide hardware standards is also hard because different workers' mobile needs require different devices.

Multiple vendor management. You can minimize both TCO and management headaches if you appoint, hire or sub-contract someone to deal with the various vendors and service providers involved with most mobile implementations. Planning, maintaining technology compatibility, enforcing project delivery times and billing are some of the business functions in which potential hidden costs are multiplied mani-fold when you work with different providers simultaneously.

Having a good systems integrator can do a lot to remove the headaches of managing multiple vendor relationships, but you need someone inside your organiza-tion who has an active role in this process. Otherwise you risk losing control of the deployment project, and increase the odds that vendors will not be kept abreast of changes within the organization that can impact the use of these applications.

Good vendor partners keep TCO low

Once you find vendors that can supply what you're looking for, talk to their government customers who are implementing similar applications as you to get feed-back on the pros and cons on the vendors' products. References for good vendors are extremely valuable.

Whether your government organization is large or small, there are some ben-efits of going with smaller vendors to meet some of your technology requirements, though this tends to run counter to the cautious nature of some IT people. With small companies you often get more attention from people who will work closely with you to provide the right solution because your contract is worth more to their growth than vendors that have hundreds of customers.

Be leery of a lot of forward pitching from large vendors, meaning they tell you "yes, we can do everything you want" but fail to mention that it will be a year later when version 3.0 ships before they can do it. Find out what's usable now and see how it can evolve over the next five years. How will it dovetail into new technology that you might buy later?

FMC Airport Systems' Don Pohly believes it's very important to find a known reputable supplier that will partner with you to evolve your application in a long-term relationship. "This type of supplier will sit down and develop a growth plan and help estimate the up-front costs, then proceed with a pilot program so you can run tests to see if the application works as advertised.

No one size fits all

Pohly suggests that working with an ASP might be a good idea for the short-term if you're still not certain if a particular mobile application is what you need, or you want the flexibility that an ASP relationship provides. For smaller government departments or agencies where even a modest investment in wireless applications is significant and they don't have a lot of IT resources, ASPs are also good partners to have.

Organizations may not care so much about the category of vendor as the quality offered. Systems Supervisor Joe Whatley of HealthPartners, a major health care provider, says "I stick by my tried and true mantra – service, service, service. If the product does what's advertised and their post-sales support meets their pre-sales advertising, then I'm happy. When I call our vendor I get an immediate answer, and they seem to be proactive with new applications. This level of service is every bit as important as the quality of the products a vendor sells."

A factor to consider when selecting vendors is that one vendor, one software platform, one type of mobile hardware probably won't meet all of the needs within your municipal or county organization if you have more than a few dozen people. This means that you will have to deal with mixing and matching applications to meet diverse needs. Duncan Bradley, Manager of Market Knowledge for RIM, states "there is no single device that satisfies everyone. Some people only need a PDA, others only need a phone. Someone else needs e-mail and data base access. You want to minimize vendors, but you will have tradeoffs."

Maybe a department that lives and dies by graphics and color-coded mechanical drawings needs color, but a person doing inventory look up and asset management doesn't. ROI will be higher if you are delivering a compelling solution to your users, so don't go cheap and try to force devices on people just because they're cheap. Give people what they need to do a good job. You will find that great Philadelphian, Benjamin Franklin, to be wise when he cautioned against being penny-wise and pound-foolish. The few pennies you save by trying to make due rather than giving your workforce what they need from the most appropriate vendor can cost quite a few dollars in replacement and re-deployment costs.

You have to plan for vendor bankruptcy or acquisition. I don't think you can bank on everyone being here in three years, at least not under their present form and ownership. Create a backup plan for every vendor so that you have a quick response and minimize the grief and aggravation in case one of them isn't there when you wake up one morning. Some vendors don't fail, they just get swallowed up by bigger companies. Then you have to worry about questions such as what if the quality of service slip, or the new owners won't support the products you already bought.

The interface that launched a 1000 ships – the value of good user interface UI design

You don't have to be first to be successful, but it definitely helps to have a good interface. The Apple Newton was supposed to be the mobile device that would launch the personal computing world into a new era. Instead, it sputtered and flamed out, due in no small measure to interface capabilities that were described by the few who used it as over-hyped and underdeveloped relative to consumer's needs.

But in 1996, the Palm Pilot 1000 launched and proceeded to take the world by storm. There's one important lesson to take from their ascendancy to dominance. The success or failure of controlling TCO depends as much on the ease with which people can operate the software user interface (UI) as it depends on the value delivered by the applications. If people can't or won't use an application because of a poorly designed UI, it will cost a lot to fix the problem.

When you implement wireless applications, set aside time and resources to make sure that you do this part of the implementation right if you're in-house staff is designing the software, or shop carefully if the vendor is responsible for the UI. Many department managers like software that has a customizable UI so they adjust it to conform to the way their employees work. This is very much a business-side issue because the UI must facilitate the way your people work.

Operational issues that affect UI design

First and foremost, focus on simplifying the data that gets delivered to mobile devices. Regardless of the application, people must be able to easily see the information they need even in bad lighting, have a minimal number of screens to navigate and input data with a minimal number of screen taps or graffiti writing. Design the UI to address the work environment.

Making your existing applications wireless-enabled can be a bigger challenge if your mobile workers are using PDAs. If you're on a desktop accessing an application, you can see everything. Call it 100 points of light. If you're on a mobile device, you go from 100 points of light down to 5 points. You have to decide what the most important information for PDAs to access is. Then prioritize the type of data that you will make available.

Second, put as much functionality on the mobile device as it can handle, particularly for emergency first responders and emergency repair crews. Even though you have fast municipal broadband, computing processes are always faster if workers complete the tasks locally and then do data uploads or downloads.

Third, test today, test tomorrow, test forever to understand not just the UI design but how people's use of the applications changes. Even with pilot tests, once you get into full deployment, users come back with issues that are 180 degrees different from what vendors and managers think the needs and problems are. People using the same device in the same city have different experiences and possibly different expectations depending on time of day and location from where they're using the device.

Fourth, do yourself a huge favor and buy applications or use ASPs that automatically take your content, filter it and format it for the various mobile devices. You have enough work to do making your applications functional for everyone who has to use them without the workload and headaches of trying to match UIs, content format, etc. for various devices that are constantly evolving.

Specific UI design recommendations

In order to articulate forcefully and accurately what you need for an interface designs requires that you know some of the basics of this technical process. A few years ago, Michael Stokowski worked at Geoworks on the team that developed user interfaces for some of the early PDAs and smartphones. He offers some thoughts about UI design for applications that run on mobile devices, and what are some key rules business managers should keep in mind.

"UI design, often overlooked in the application design process, is as important as the software coding itself, especially in wireless applications," Stokowski remarks. "The application obviously must solve the problems the user expects it to solve, but the UI must make the solutions easily accessible. For instance, a data collection tool must make it easy to enter contact information, not just store that information in an elegant database for future use."

Clearly define a UI design process, and hire or contract the right people to do the work. If you hire the right people, they can help define the design process. Here is a summary of Stokowski's key points in the UI design process.

1. Identify the user requirements

2. Define the feature set

3. Conceptualize the UI

4. Design the UI

5. Layout the UI

6. Test how well people can operate the UI

7. Fix the problems

8. Code the application

To sum it up

Figuring out ahead of time what your total cost of ownership is going to be is 50% analytics, 50% crystal ball gazing, particularly as it pertains to the broadband network. However you approach the mechanics of the task, it is important that you allow for more of everything. More time, more money, more headaches, more odds and ends. Don't go overboard, of course, but give yourself some room for Murphy's Law.

Press everyone you work with inside and outside of the organization to take a second and third look at all of the factors that drive costs. Cities, and government organizations of all types, are great at exchanging experiences and ideas. Use this to your advantage and find out what other governments are doing, what expenses they're encountering and how they are keeping costs contained. Go to your business communities that have deployed wireless applications and see how they have managed TCO.

TCO must not be an afterthought with your organization. That said, do not let people use TCO as an excuse to delay moving forward. "How do we know such and such won't cost more than estimated?" is a valid question. But as Robert Bright says, "sometimes you just can't tell what's going to happen until you move forward." Put someone in charge of monitoring deployments for those factors that can increase costs – delays, equipment breakage, financial news, political activities, etc. As they spot something significant, make adjustments and move forward.

Chapter 12

The ROI at the End of the Rainbow

The most satisfying element of mobile and wireless implementations is taking time after a year to assess the impact of the application on your organization. This is always a positive experience because if done correctly, you can learn a great deal about the technology, your organization and your implementation processes. As important as assessing the ROI impact of your deployment is uncovering where you can improve the benefits you receive during the second year after your deployment.

How organizations go about this post-game analysis varies depending on the type and complexity of the application, the size of deployment and many factors unique to the organization itself. But there are some basic ground rules you can adapt and follow to meet your needs. Their effectiveness depends on how well you know what you're looking for, and how honest the organization is with itself interpreting what it finds.

Evaluating the broadband network's ROI is the greater challenge compared to quantifying ROI for mobile applications given that much of the benefits derived from the infrastructure depends on how the departments and constituent groups use it. Work with these groups to establish ROI objectives before the pilot, then review them right after the pilots as discussed in Chapter 10.

What everyone should come to consensus on is, 1) are the objectives realistic, and 2) should the ROI objectives be more clearly defined. When I look at the objectives that cities present in their RFPs or RFIs, these are easy to comprehend, ambitious, noble and inspire support. But are they realistic? Does the city have the resources, and more importantly do they have the will, to undertake and complete all the post-deployment work that's required to get the job done? Do these objectives, or at least the objectives internally communicated within the government, have enough specific detail so everyone unequivocally knows when they have achieved something worthwhile?

The previous chapters gave you a high-level view of the many tasks, challenges and realities associated with using wireless technology to change the way local governments do business. You have to decide whether your ROI objectives are realistic in light of what you see facing you on the road ahead. There's no formula I can think of to help you here.

This last chapter helps you review the factors that are involved with establishing some criteria so your ROI analysis is worthwhile. It points back to the framework I presented in Chapter 2 for developing strategic and tactical objectives, which serves as the foundation for the ROI objectives you set, and gives you some recommendations on how to proceed. Like most processes, you get out of it what you put into it.

Looking for ROI in all the right places

John Dolmetsch has been involved with a million square miles of municipal deployment of one type of wireless infrastructure or another. He points right away to the low hanging fruit in terms of making the ROI case. In most areas of the country there's a significant cost justification to deploy these networks outside of generating wireless access subscribers. Unless they have fiber linking their buildings together, some cities' ROI model could justify the investment in broadband WiFi within 24 months through its impact on replacing the recurring fees they're paying for inter-building connectivity such as T1 lines.

"If you look at the city of Philadelphia, they could probably cost justify building a network just to connect its own schools and government building, achieving ROI somewhere between 18 and 36 months. I haven't seen one city yet where that's not the case. Typically when we go into an area and we start dealing with a government, the first thing we tell them is to go back and add up all your current costs for connecting buildings and users within those buildings."

His is a variation on my premise that the cost benefits of wireless-enabling your government operations and mobile workforce can justify whatever else you do with the network. From John's viewpoint, building the network to eliminate recurring costs lays the groundwork for the rest of the infrastructure. To use his analogy, "you build main highway routes through your area. Once these are in place, building the sides streets are nowhere near as expensive. Carrying that over to wireless, after you have the basic network infrastructure in place, adding one more building or giving 150 people mobile devices is really not going to significantly impact the cost of the network." These extra benefits you get are bonus bucks.

Going back to the foundation

To expand the ROI analysis process, let's look back at the framework I presented in Chapter 2 to help you build the foundation of your business case for wireless. The four main strategic areas were: 1) communicating more effectively and efficiently with existing constituents; 2) enabling constituents to get more effective service and support; 3) improving internal communications and business operations and 4) communicating more effectively with potential constituents.

Do a thorough analysis of how departments and constituent organizations operate in these strategic areas, assign quantitative values to tasks performed in these areas and calculate the financial impact of the network improving or eliminating these tasks. Determining ROI is the relatively straightforward exercise of calculating deployment and management costs plus whatever else contributes to TCO, then subtracting this amount from the various financial benefits the network actually achieves. Take note, however, that this is simple to describe, but it can be a tedious, complex exercise as you bring all of the departments into the big picture.

To put this is context, let's look at the hypothetical example I gave in Chapter 2 of how a city could set a goal to save $300,000 in annual communication costs for distributing information to low-income communities. Keeping things simple, assume that the cost for deploying the network in that community was $300,000 (the actual cost a town of 6,000 paid to deploy their network).

The first year's TCO when you factor in maintaining the network, getting donated PCs into hands of residents, training programs, content development for a community portal and other related expenses brings the cost of the project to $400,000. After one year of the network's operation, an analysis of the actual reduction in communication costs shows that the city is only saving $200,000. Therefore it will take a little over two years to pay off the initial $400,000 investment and start seeing a positive ROI.

These numbers are purely hypothetical and simplistic to give you a general idea of how to step through an ROI analysis. You finance director and an accounting firm can give you a definitive picture of how to calculate your actual situation given the specifics of your municipality, nature of the deployment you're planning, etc.

Measuring ROI

Constituent portals have analysis tools to help you measure how much of what kind of information is being pushed out to, or accessed by constituents. These plus internal analyses can quantify the impact of the network on cities' business and marketing communication budgets. Web analysis and traditional survey tools can tell you what services citizens and visitors are receiving through the network and their level of satisfaction. Here too you can measure the technology's impact on your service delivery operation.

There are various ways to measure improvements in internal operations once the network launches, starting with how much you're saving by eliminating recurring cellular data service charges. If you wireless-enable many of your assets in addition to your workforce, the cost-savings and equipment performance enhancement due to improved asset maintenance alone can be a real eye-opener within a year or so.

Set up achievement milestones for three, six and nine months following the launch, and see how well you reach or surpass these milestones. It not only builds your political capital for other wireless initiatives when you're successful, but if you're not quite making the numbers for one quarter, you can make mid-course adjustments. Since large metropolitan areas will be deployed in segments, you can start tracking results for just those neighborhoods as they come online, though the full ROI likely won't come until the entire network is up and running.

How and how often you report your ROI details depends on your particular circumstances. You might want to be prepared to do quarterly reporting if the original political heat opposing the initiative was high, or you have some other initiatives you want to fast track through the budget approval process before the year is up. Otherwise, semi-annual or annual reporting should be fine. Internally, though, monitor these numbers like a hawk during the next year until you have other city or county deployments to serve as benchmarks for your milestones. Keep your vendors appraised in case technical factors on their side are influencing ROI. Once you have benchmarks from other cities, determining if your ROI is "normal" is a little easier.

Intangibles make ROI analysis difficult

Part of the answer to the question "Did we get our money's worth?" is determined by dollars earned or saved, time saved, in increase in tasks performed with the same number of staff and other quantifiables. However, the network produces intangible benefits, as Robert Bright will tell you, that make a formulaic approach to analyzing ROI inappropriate at times.

"Are you trying to maximize profits and city dollars?" he asks. "That's part of equation, but that's not the whole situation. You can't put a price on increasing school education levels of the population, on improving access to educational opportunities. It's very difficult to put an absolute number on many things a city does, part of this dollar goes here, part of that dollar goes there. We as a community have to step back and recognize this when we're talking about what services to offer, how we price it.

You can decide how the intangible benefits of the network can be evaluated to determine an acceptable ROI through the consensus-building and expectation management recommended in previous chapters. During the Philadelphia focus groups and town meetings most of the participants agreed that economic development and closing the digital divide were two areas where they expect to find the bulk of the ROI as long as tax dollars weren't spent in the process.

Given that this is an apparent general consensus of many of the city's citizens, is the initiative on its way to a positive return on investment? Well, given that Wireless Philadelphia is a fully operational non-profit entity with the mission to use the network to attack economic and digital divide issues, they have identified a revenue source and Earthlink is footing the bill for deployment, I'd say yes. Definitely. Of course, there is a lot of organization management work left to be done and people's time is money, but odds are with Philly. Remember, there's also the estimated $1 - $2 million in annual savings for wireless service charges that the city's 2,000 mobile workers won't incur once they start using the network. Those are serious bonus bucks.

So all of this hand-wringing and report-generating "pious opposition" (as Ed Swartz describes it) stating that the city can't generate its projected subscription numbers for the service is, quoting Shakespeare, "full of sound and fury, signifying nothing." That's not the main ROI objective for this network. Even if it was, I believe that once local businesses realize they can have all of their mobile-about-town executives and employees equipped with high-speed wireless access for $20/ per month, it's "Katie, bar the door" time. There's going to be a major rush on service signups. And what about when businesses figure out they can buy into the network rather than build their own WiFi networks for outdoor facilities and warehouses? Subscription sales should be just fine, thank you very much.

Evaluating the ROI of workforce applications

Project team leaders for small workforces or individual departments have a clear-cut job when analyzing the ROI of their applications. The user community is small, the application is probably simple in terms of its objectives and the results are easy to track. Maybe you can calculate ROI on the back of a napkin.

In other deployments, the feeling of job improvement among mobile administrators or workers can be so great that it doesn't matter if you have numbers to support an ROI or not. People will riot before giving up their mobile devices. Wireless access to e-mail, very difficult to analyze for ROI, creates fanatical devotion among its many users. It's not without good reason that RIM BlackBerry devices are referred to as CrackBerrys.

For some departments, as is the case with Philadelphia's L & I, the application tackles specific sets of job functions and it's clear when the application is achieving benefits that justify the expense. Then, when they determine that there are other benefits, such as mobile workers being able to link into applications and databases of other departments or state and federal agencies, these benefits become the "bonus bucks" of the deployment.

However, in some cases the ROI analysis may not produce clear-cut or expected results. Benefits that were unforeseen at the start of the project can pop up in individual and organization-wide improvements if you probe deeply enough to find them. Or conversely, some of the promised ROI falls short of expectations because of factors beyond project teams' control. There are frequently intangible benefits wireless-enabling some workers because the bottom line is still that governments deliver community services for which it's difficult or impossible to set a price.

Sometimes, it's a challenge just to be able to complete an ROI analysis. One major barrier to good analysis in some organizations is that managers are forever busy and needing to move on to the next project with barely time to catch their breath. So these folks are hard pressed to find time to be thorough in their review.

They do a quick check with their staff, make sure there aren't technical glitches holding people back and if it's clear that productivity is better than it was before the application, everyone is happy.

A barrier that some people don't like to discuss is that the team didn't clearly define before the project started what its objectives were. It's hard to know if you're "there" yet when no one agreed on where "there" is. Another barrier is the lack of vision, or depth of vision, by those doing the analysis. They have a myopic view of ROI that blinds them to the big picture, maybe one that focuses too much on quantity of service calls processed rather than quality of service delivered.

Did you achieve the planned ROI?

Here are some general tips to help you evaluate how far you've come. 1) Focus on one or two of the most important benefits the application can deliver, and then look carefully at how well those benefits were delivered. 2) During ROI analysis, actively search for the un-planned benefits since these could be more beneficial to an organization than the original reasons for buying the application. 3) Open everyone's mind to explore the future benefits once workers get comfortable and truly proficient with the technology, which is advisable if a deployment is particularly difficult and some workers lose faith early on in the application.

The "single purpose in being" approach has merit. You can look at Scottsburg IN, for example, says Jeff Arnold from the National Association of Counties. "They've done a great job. They were able to keep three plants that are main sources of jobs in that area because they stepped up and spent almost $300,000 to set up a strong wireless network. But they had that critical motivator. The companies said 'either we have broadband connectivity or we're out of here.' That was the incentive for Scottsburg to move forward instantly and try to find a solution that made sense. The wireless solution for them certainly was the way to go in terms of benefits relative to the costs."

If you have a manufacturing plant or a wheat processing facility that needs high speed connectivity, then the benefits to the community of improving their business operations plus the resulting increase of tax revenues might make sense. Along with their main objective Scottsburg also created a business model in which they specified what turned out to be an obtainable number of paid access subscriptions to sell in the first year that would offset the investment.

This focus on one or two main objectives theoretically leads to the easiest and fastest analysis to do. But be realistic in just how much dollar value you place on some benefits. People can end up spending more time on this than is necessary, such as those who went through extensive contortions trying to tie the time saved using wireless e-mail to specific dollar amounts. If users save 11 hours processing e-mail, how do you know they will convert that time into productive work? Is that big deal the city manager saved because she was able to negotiate the terms instantly by e-mail from the airport truly indicative of the dollar value of the application? Sometimes it's enough just to know that people are saving 11 hours.

Determine before you begin the pilot project how much probing and calculating you're going to do to assign dollar values to your ROI, or if your analysis primarily will be one to put numbers into a business context. The ROI analysis of the latter type might conclude that "our utility repair people aren't sitting around the office an hour in the morning waiting for dispatch assignments, and subsequently they're putting more hours into customer service."

If the financial-impact numbers are going to be an important element, then don't cut corners on the analysis. Be exacting in how you set the criteria for quantifying and measuring success after deployment. Make sure everyone's comparing apples to apples. For example, if productivity gains are to be measured by time savings on each constituent visit, make sure all of the departments are assigning a value per employee in the same way, such as salary plus "a," "b" and "c" benefits.

Numbers speak volumes when it's time to set budgets, but some in senior management may be content with anecdotal evidence of improvements. If you cannot quantify improvements in specific numbers, look for specific changes that benefit the organization. Maybe inspectors end up not having to ask construction contractors questions since the data is already available on their handhelds, so they spend more time offering advice to owners on how they can get their jobs for the city completed sooner. One of the big values of wireless and mobile applications offer is the potential to change how organizations operate. When you can, anticipate what these changes may be and factor them into the ROI analysis.

In many of Dianah's presentations, she makes effective use of anecdotes about specific individuals whose lives were changed through programs enabled by wireless. The teenagers who are now helping people in the community use computers. The handicapped mom who started a business in her home. The same type of stories of workplace events that paint compelling pictures of government helping people thanks to technology packs an ROI punch.

The unplanned benefits may be the greater ROI

While job security is definitely enhanced if you deliver on the expected benefits, career advancement certainly gets a boost if you set your analytical sites on uncovering benefits others didn't foresee. To do this, collect a lot of feedback from end users and take a hard look at the data that is coming into the organization as a result of the application.

When workers initially hear about a particular application, some of them can comprehend a few of the intended benefits. But it's not until people actually use the technology in their jobs do they fully understand its potential. Smart organizations encourage users to bring ideas to management for improving business operations with the technology. Smarter organizations aggressively probe in depth to discover what users are doing that they weren't able to do before deployment so managers can factor these benefits into the ROI analysis and plan for additional information.

Now that social workers are carrying laptops to help them process clients' paperwork for city services, the intended benfits, can they also tie into county or state agencies databases to help clients get issues resolved with these agencies during the same visit? Can park workers with devices that connect to resources that help them direct tourists and visitors to popular destinations also track vehicles, maintenance equipment and other assets?

Administrators who get wireless access to e-mail realize they can improve group scheduling with city contractors and consultants, a possibility that project planners had overlooked. Once they start using a work order dispatch application, work crews realize that wireless access to the scheduler enables them to schedule preventative and reactive maintenance from the field. The greater the unexpected ROI, the better should be your leverage for getting budget for the next application. So ask the right questions. Leave no stone unturned when gathering feedback from those who know best, the people in the trenches doing the work.

Critically assess what you can do with all of the data a wireless application brings into the organization faster and in greater volume. Knowledge is power. Knowledge gathered in real time and properly applied produces increased constituent satisfaction and more efficient operations. Few things frustrate constituents more than sitting in meetings with city employees who can't give them answers to important questions because the answers lie in some other department. Don't over-look the fact that data itself is an asset that increases in value when all of the city's data sources are consolidated.

Quantify the impact wireless and mobile have on improving the ROI of your other software applications. For example, managers may not immediately see the impact that giving first responders mobile devices can have on research analysis software the public health department is using to track health trends. But data, once it's captured on network servers, is easily transferable to any number of applications.

Instant delivery of job-completion data from field inspectors speeds up the billing software's ability to process invoices for violation fees. RFID and warehouse management data captured by workers' mobile devices can improve the effective-ness of fire department workers during an emergency. Bring representatives from all the departments that touch, or are influenced by data that mobile workers collect so you can determine how the ROI of their various applications subsequently are increasing.

The future factors into ROI

If your IT staff builds a software architecture that facilitates technology inte-gration, then you have a potential stepping off point for numerous applications to add to your initial wireless deployment. This translates into financial benefits is through time and money savings.

Once the initial application is delivering benefits that everyone buys into, you don't have to spend a lot of time selling management and end users on the next applications. People are coming to you with ideas. From a development standpoint, IT has an easier time building follow-on applications because a lot of the heavy lifting is done on the initial program. Backend server enhancements, security proce-dures and hardware and software evaluations should require only rudimentary resource commitments for future projects.

To get a feel for the extent to which you can plan "down the road" ROI potential, use feedback-gathering interviews with end users to also talk about future applications. Of course, don't overdo things with outlandish pie-in-the-sky numbers. Focus projections on applications that can be developed or expanded upon within the next year or two, while giving secondary considerations to applications and technologies that might be three-to-five-years out. As you complete the shorter-term projects, refine your plans as well as estimates on the longer-term applications.

Conduct one or two brainstorming sessions with senior level officials and line managers to ponder possible next ventures. This level of buy-in for future plans is helpful if the ROI numbers for the initial implementation are good, but not outstanding. People who can see the big picture view tend to be forgiving if the short term isn't as great as everyone hoped, as long as these visionaries believe the project still has plenty of upside potential.

Budget management issues

In the last chapter I talked about how TCO impacts ROI: the greater the TCO, the harder it is to generate a significant return. Along with accurately predicting and controlling TCO, another factor impacting ROI is how well you manage to stay within your budget once the deployment gets started.

Needless to say, accurately calculating your budget right after completing the pilot helps avoid cost overruns. It's relatively easier to do with mobile applications than the network infrastructure because prices for the component parts are easier to predict and lockdown. Given the reality that every network infrastructure deployment large or small creates a unique selection of equipment, quantities of components, configuration, topography challenges and political unknowns, budgets can be fluid. That said, there are some factors to consider.

Budget setting for infrastructure deployments

To try to put at least a little predictability into the infrastructure budgeting process, I spent some time with Ed Taulbee, Director of Carrier Markets at Tropos Networks to get an idea of what deploying the WiFi network infrastructure might cost. Topography is a major determinant of cost because the lay of the land determines how many access points, gateways and other component parts you need. He breaks topography into three types: 1) dense foliage or urban canyon, 2) greater metropolitan suburban and 3) low obstruction.

Many municipalities and definitely a lot of counties are comprised of a combination of these types. The dense foliage/urban canyon topography are the areas where you have lots of tall buildings, many neighborhoods with houses located close to each other and parks packed with trees and hills. Downtown Philly, Manhattan, NY and much of San Francisco are the typical places falling into the first category.

Northeastern bedroom communities such as Long Island and other suburban New York City areas, suburban Philly, and many of the mid-sized cities fall into the second category of greater metropolitan suburban. They may have a fair number of tall buildings, and a relatively large population, but not nearly as densely packed as the big city downtown areas. There may be more trees, but these are interspersed with fewer people and buildings.

Mid-western and western cities such as Phoenix, and AZ, Oklahoma City, towns and all rural areas fall into the last category. There aren't a lot of very tall buildings, everything and everybody are more spread out and you may not have a lot of vertical assets on which to mount network equipment. Many of the new housing development areas in suburbs and mid-size communities fall into the low obstruction category given that the hills are frequently leveled or significantly reduced, commercial buildings are of the mall/strip mall variety and trees are thinned.

Building the infrastructure in a dense foliage or urban canyon area can cost between $125,000 and $140,000 per square mile. The greater metro suburban areas can cost about $95,000 per square mile, and the low obstruction areas may cost about $50,000 per square mile. So your first step in determining budget is to bring in a consultant such as Civitium or MRI or a systems integrator with solid experience estimating these types of projects, and analyze the area of planned deployment. How much of it fits into which category? In a city such as Philly, maybe 20% of it is urban canyon while in Tempe, AZ you may find 0% in this category.

Additional factors impact the budget

There are several other factors to consider, though. "These cost estimates represent everything that's going into the network deployment - radios, gateways, backhaul and the back-office operation with servers, routers, switches for authentication, operations support, bandwidth shaping and the like," states Taulbee. "But the final total really depends on what performance are you willing to drive and at what costs." If you have a port area where you are using the technology to support a small number of workers and asset management tasks that don't transmit a lot of data, you're not going to overpopulate this area with access points. But if the network has to support an area with 100 or 250 users per square block, you'll need a higher density of access points.

"Are you maximizing the value of asset re-utilization?" is an important question that Taulbee asks. "Typically cities have fiber networks coming in that they built or are leasing, so use those for the backhaul. Also, rather than build a network operating center (NOC), which is expensive for business grade service since you need people in there working 24/7, a larger city may use a call center that its utility operation already has in place. Just add several people to the utility's staff who primarily manage network infrastructure operations, but who can do other tasks for the call center during times when network customer service calls and usage are minimal."

For communities that initially plan a limited initial downtown deployment or a single-use application, but may expand the network later (i.e. Houston's parking meter operation), consider over-building the back-office operations center. Whereas the network data performance is directly related to the infrastructure of radios, gateways and the backhaul, the back office is an expensive element of the build out and then upgrade. Putting more capacity into the back office for the short term better controls your budget in the long run.

Whether you use license and unlicensed spectrum is going to have a budget impact. Wireless Philadelphia Project Manager Varinia Robinson comments that "If you're planning to deploy some form of WiFi, that's unlicensed spectrum and it doesn't cost anything to use. But if your project team decides that they want buy licensed spectrum, such as what you have with networks built specifically for public safety use by police and other emergency first responders, there are extra costs associated with this. These networks require proprietary equipment, both on the network and what homes and offices use for customer premise equipment (CPE)."

Earlier I cited remarks from Civitium about the cost of CPEs and how this affects the end user's budget. WiFi works on the CPEs that come built in with most laptops. Often indoor users need a bridge, but these are being supplied in some cases by the service providers when customers initially subscribe because the enhanced performance eliminates a lot of expensive customer support by the provider. In rural settings where typical radios aren't practical, CPEs need to be more powerful and therefore cost more. This is going to impact the budget of someone along the line, whether the government, the service providers or the end users.

Budgeting for mobile applications

It's somewhat easier to do budget planning for mobile applications because prices for the component parts are easier to predict and lockdown. However, the longer the pilot development and testing period and the greater the number of users to whom you are deploying, the greater are the odds that market prices and particular product lines will change before the year or so it takes to go from pilot to the completion of full deployment. If you can, try to factor this change into the contract because in a year, the product you order today likely will cost less and a new product coming out then that you might want instead likely will cost more.

When establishing a budget from scratch in an area with few benchmarks, it helps to practice a little restraint. There's no need to buy $2500 laptops when $400 smartphones will do the job. This is stating the obvious, but if you're solving a set of simple problems, use economic off-the-shelf products and save the big technology guns for resolving complex needs.

As a general budget guestimate, you may be able to equip 1000 people to wirelessly access several basic back office applications such as inspections management or data collection $600,000. $500,000 of that is for handheld mobile devices and training, $20,000 is for servers and about $80,000 for other related middleware costs. This will go up significantly if you need greater processing power on the device, or the ability to view and manipulate more data on the screen. Then you're probably in the range of $1,400 per tablet PC type of device when bought in huge quantities.

Since the broadband network enables high speed access for any device with WiFi capabilities, you don't have to automatically replace existing laptops with PDAs and smartphones. Mobile workers who operate under similar conditions as they do in the office (i.e. sitting down) and have time for their machines to boot up before working with a constituent, can keep their laptops. It saves you money for devices and mobile device management expenses.

Don't forget to factor in the cost of development time for your IT staff to customize mobile applications and provide end user support. If the vendor does all of the customizing work, sometimes they factor this into the price of the applications while other vendors bill this as a separate cost. And always factor in TCO.

How small organizations and non-profits can approach budgeting

The previous costs I outlined are for large city or county governments. Smaller organizations still have the per-person costs of mobile devices, but since they have smaller workforces their implementation costs can be much less, particularly if they use a wireless ASP. For example, a simple wireless dispatching system for animal control workers may cost about $50 - $75 per person each month. Set up charges may cost $1000 or so depending on the scope of the wireless effort.

Non-profit constituent groups tackling economic development and digital divide issues may be able to extend the buying power of their budgets by getting corporate donors to underwrite the costs for the technology projects. Second Harvest, for example, feeds homeless with reclaimed food. An organization such as theirs can look to industry to donate products by showing that if the delivery people had mobile devices, they could do more to increase food collection and facilitate food distribution.

However, requesting vendor donations as part of your strategy to get technology might be easy if you request products at the end of their life cycles. The greater challenge is getting application development and deployment services donated. The key to securing these services may be found in the fact that more donors are demanding that non-profits use better business practices. To comply, non-profits can seek in-kind donations by executives of wireless product or services companies who join their boards of directors or planning committees to bring application development expertise to the organizations.

Depending on the size of the organization, the day-to-day operational needs of keeping the technology up and running are another budget item. The People's Emergency Center in Philadelphia is taking the creative approach of teaching neighborhood teens how to build, use and repair computers, then having them provide tech support services to others in the community. Other organizations should consider similar tactics for making their money go further.

To sum it up

When all is said and done, your ability to measure the return on your investment for wireless infrastructure and applications that change how your government does business depends on two things: how clear you are about where you're going, and how well you keep track of results as you get there. The process of doing both of these things can be simple or exceedingly complex. There is no universal right or wrong way, though if your city has a good track record implementing technology applications in time and within budget, it's probably good advice to keep those practices in place for wireless.

Be sure to get all of your officials and department managers, stakeholders, constituent groups and even the media on the same page when it comes to your ROI objectives and how these will be quantified after the technology is in place. Maintain a certain amount of flexibility in the ROI analysis process. Use your pilot projects to test and fine-tune ROI assumptions.

Probably the hardest part of any technology deployment comes if someone has to say "you know, try as hard as we may, we can't get there from here." There are times when the ROI everyone expected or wanted can't be attained with the products, vendors, approach or some other factors currently in place or on the planning board. The politics alone could make this kind of a pronouncement unpalatable. Hopefully someone can make the hard decisions to change what needs changing. Or you adjust people's expectations as you adjust the ROI objectives.

The important thing here, as with everything else discussed in this book, is that your make decisions which are best for your particular city or county. When you can, rely on other governments and their experiences as guideposts. But ultimately, do what works best for your area.

The following epilogue brings together various people involved with the Wireless Philadelphia initiative or other municipal broadband projects, to offer some closing thoughts as you move forward with your initiative.

Epilogue

Looking Forward

Many people directly and indirectly involved with making Philadelphia a digital city with high speed broadband wireless access available everywhere are, as you can imagine, passionate in their support of the project. They believe the impact from this effort will have depth within the city and breadth well beyond the city, probably throughout much of the world.

Philadelphia is on the eve of launching its full deployment of the WiFi network. A huge amount of work has gone into bringing Wireless Philadelphia this far. A lot more still lies before them, Earthlink, its vendor partners, the various constituent organizations, and the many citizens who will be the technology's direct beneficiaries. Here are some summary thoughts from a number of those individuals as they ponder what the future holds.

How's it look from the top?

The Hon. John F. Street, Mayor – City of Philadelphia

Understand that a wireless solution potentially will be different for each city based on their communities' needs. There are a number of business models that each city will need to evaluate based on their own criteria.

In the final analysis, do your homework and get community involvement. Keep the process open, public, competitive, and fair. Have a careful, well-considered strategy for dealing with the myriad of financial, logistical and political issues that will undoubtedly arise. True change requires political courage, and a willingness to see beyond the daily crises of government to look into the future, envision new realities and then make those realities happen.

Philadelphia has shown that you can be creative in your approach and ensure there is open competition to bring economic stimulus to your community. As a result, the citizens of this great city are poised to be some of the most tech savvy citizens of the United States.

Dianah Neff, CIO – City of Philadelphia

We think that what we've projected in our business plan is coming through, and we're very excited about this initiative. Looking at the impact on tourism alone, we get 25 million visitors annually. Think about the 1% that want access to wireless anywhere they go in our community and the potential that that market stream can bring. Think about what this means to your municipality, particularly as more cities get involved and visitors get used to that access. It helps your restaurants and entertainment establishments.

Sometimes it gets a little frustrating with the delays and roadblocks people keep putting in our way. But then I look at how much we've changed the market and I realize when you're creating revolutionary change, this really hasn't taken so much time. It is time, though, for municipalities to step forward.

I firmly believe that every city needs a strategy, even if that strategy is 'we don't need municipal WiFi at this time, but we will continue to do the due diligence as the technology evolves.'

Varinia Robinson, Project Manager- Wireless Philadelphia

In two years I expect there to be a much higher percentage of people who have broadband access and can use the Internet to their advantage. The local business economy and the quality of life for people will improve as this technology helps everyone across the board. We'll definitely see it help schools with their outreach to students and parents. There are so many ways it can enhance tourism as well. People carrying handhelds with WiFi capabilities, for example, could go to the Liberty Bell and push a button on screen to hear information about it. This is a transforming technology. Cities have a responsibility to bring this to their communities to empower the people.

Robert Solvibile, Acting Commissioner, Department of Licenses & Inspections – City of Philadelphia

10 years ago, cell phones were big, they looked like walkie-talkies and only a few of the elite had them. Now little grade school kids are walking to school with cell phones on their heads. I see computers and wireless doing that for the citizens of Philly where it will be convenient for everyone to have that level of access. In time people will probably be able to get a $100 computer that they use for three years and then they go buy their next $100 computer. I see wireless being a boon to the city. There'll be all kinds of applications in the schools. It's hard to imagine everything it will do, but I believe that lots of good things will come from this.

Where are we going?

Patricia Renzulli, CIO - School District of Philadelphia

When I look back on the work of the Executive Committee, I see a group of people who all came in with the desire to make this a success because we work for constituents who will benefit from the initiative. When you think about it, this is not about the network, that's just the delivery mechanism. It's all about the content, whether that's a business transaction, delivery of services to small businesses, or delivery of instructional materials to students and parents by educators. When you look at all of the potential services we're talking about here, they all sit on top of the infrastructure. We're not able to deliver any of that if we don't have the infrastructure.

Paul Butcher, Marketing Manager, State and Local Governments, America's Marketing Group – Intel, Corp

Municipal wireless is the right thing to do. The bottom line is that our city and state level governments are under incredible financial strain. Government can be more efficient and effective in generating revenue streams, and cost outflows can be re-directed more efficiently. Using wireless for job reengineering generates savings which translate into cash flow improvements that are part of a sustainable business model for cities working in partnership with the right vendors and service providers.

There are hundreds of communities that are starting this move to municipal WiFi or looking at doing it. Not only Philadelphia, San Francisco, Portland and Houston, but 20 different communities in Florida, Michigan, Minnesota. This is very real internationally. The motivations are universal. It's a market force that is happening, it's a paradigm shift. I think in major metropolitan areas the trend is to move from the utility model to one in which partnership consortiums are the service providers.

I feel that statewide and federal legislatures are hard pressed to comprehend all the complexities and the myriad of needs that communities have. So there shouldn't be any state laws limiting their activities. At the same time, communities need to continue going about this in a transparent fashion using the RFP process, and enable the private sector to make suggestions and recommendations. The communities should be able to move forward in whatever way makes sense.

Patricia DeCarlo, Executive Director – Norris Square Civic Association

With wireless portals you have this local inexpensive way for people to communicate with each other. That vision, to me, has a lot of appeal. It's reinventing the way that people reach out to others. You don't know how much money is currently being spent by all kinds of organizations to inform folks about what's happening, money that could be saved with a good portal.

It's critical that you get it right about what information you're providing. If you provide irrelevant information online, then folks won't read it and they won't go to it again. I feel that in the future there's only going to be wireless. Look at New Orleans. The main way people are communicating after that disaster is through their cell phones.

Robert Bright, President - Talson Solutions, LLC, Member, Board of Directors – Wireless Philadelphia

I'm a small business owner and from my perspective, municipal broadband is going to increase the efficiency of small businesses. Without question this technology, for lack of a better description, keeps you in the game in terms of being efficient. For dealing with clients and potential clients, for making more efficient use of your time to pursue business and grow it. Like anything else, if you're not in the game, you're not going to get the call. If people don't know who you are or where you are, don't expect the call like I got 20 minutes ago asking 'here's a project, are you interested?'

As a small business, wireless helps solidify your position. How you use it, whether you give the office wireless access or your mobile employees are wireless, everybody has to make their own call. It may reduce, if you so choose, your cost of doing business. In time, you may want to make wireless a redundant communication system, so that may increase your costs. Do I need to create a dual system? I may not. But however you look at it, for small businesses that aren't using the technology, it may take them out of business.

How we're going to get there?

Rizwan Khaliq - IBM

In the dot com era anything people threw out there stuck for that moment without any real valid business processes behind it. And we saw what happened to those companies. I think there was a real danger of sound bite coverage of Philadelphia's struggles with the incumbents doing the same thing to this whole community broadband area, encouraging a business model that wasn't viable. I'm glad to see that since last year, positions and the city's business model have matured.

We understand the drivers behind the initiative, economic development, which is important. Now the scope of benefits is expanded to include the initiative's impact on the workforce. You may have hundreds of workers who can only see five or six constituents a day because of travel time, picking up paper files, all that time lost or wasted. If you can optimize their time by 10% or 15%, here's a specific financial value you can extract from the network and at the same time, transfer this into social and economic programs.

These municipal initiatives are going to result in a re-classification or expansion of what a service provider is. They will be the entity providing the network operating centers and the customer service operations. The city will bring in a wholesale model where they in essence enable these entities to sell access services for a lot less than what the telcos and cable companies can. There's going to be an increase in the number of Wireless ISPs coming out now. A lot of them will have a very slim overhead but a strong value proposition of being the coordinators of all of these services.

Everything is going to continue to evolve. The government industry is absolutely going to have a strong role to play. As the business model matures, government and service providers will end up finding their space and they will help each other evolve the model. I don't think that it will be just one or the other. Governments are in a good position to encourage industries to help them address some of pressing business and economic issues they face.

Bailey White, VP Marketing/Communications, Innovation Philadelphia

How do the cities know what's really required to make this work and how they're going to use this technology? They need to be asking the right questions of themselves before they ask questions of vendors. The answers vary. How much access is out there right now? What's your population like right now, is it a group that's hungry for wireless, is it a group that's suffering from not being part of the knowledge industry? Are you going to do case management with this, is it going to be primarily for public safety, is it going to be a consumer network?

Thoughts on fighting the good fight?

Ed Schwartz, President - Institute for the Study of Civic Values

The companies having attacked the project initially because it might succeed and therefore provide them with unacceptable levels of competition have now shifted ground entirely. They're saying that it's going to fail because the assumption of subscriber penetration rate is too ambitious and this sort of thing. I don't know what to make of this. All I know is that large companies bid on the project and didn't seem put off by that estimation.

When Ben Franklin started the library, I'm sure there were people saying you're throwing the book stores out of business. You have to ask what is in the public interest and what is in the community interest to do. If we open the door, where does it end? I don't know. However, that's not the issue. Maybe municipal WiFi is anti-*your* business, but it's very pro-business in a lot of other areas. Yours is not the only business.

Now, obviously government has enough things that it must do and it can't get involved in every business opportunity that comes along. But this is an infrastructure project aimed at creating the capacity to take advantage of a new technology, and create some ability to manage the use of the technology by a lot of people who wouldn't otherwise have access to it. If these opportunities aren't put in place due to the pious opposition of a few companies, no one's going to benefit.

Craig Newman, Director of Business Development - Motorola's Canopy Wireless Broadband business unit

The mistruths will continue until we prove it out. That's why it's so important for Philadelphia and Portland and these other early cities to get their networks up and running. And they have to get it right, or else you'll hear those people saying 'aha, I told you it wouldn't work.' However, for now the network doesn't necessarily have to make money right away, just technically operate the way people want it to. It's a new technology and some service providers are willing to wait for the ROI, maybe even two or three years.

The Hon. Glenn Sangiovanni, Mayor – St. Cloud, FL

Access isn't just an issue for low income neighborhoods. There's a gentleman who has an environmental business in which he has to take digital pictures of all of the cleanup and other things he's doing, then get these out to the state for regulatory reasons. But he can't do it from his home because the files are so big. So he comes into town every day at 3:00 to use our broadband service.

This guy's home is in a nice part of town, so affordability isn't an issue. The local phone companies looked at the community and felt there aren't enough customers for them to provide high speed out there. This is really an example of how municipal utilities got started in the first place, doing things that companies weren't interested in doing because these didn't fit into their business model.

The state legislators thought we were trying to get into the private sector's business. But that's not what we're doing. We're trying to provide a public service. I often said at the Statehouse, if you can show me a private company that wants to come in and do the same project that we're providing and at the same price, we'll step aside. But we didn't get any takers because we're not charging for it.

Yeah, they do lose that $45 a month that I won't be spending now. However, this is just the beginning. There are so many possible applications out there that could be developed for wireless. This network we're building is just putting the road out there. The telcos have many opportunities to come in with new services if they're creative enough to take advantage of the road. We're only providing a base level service. There's a market out there for the high end user, the business that wants more of this, more of that. Who knows what the possibilities are from the business sector. I think that telcos just need to get creative and look at different options.

What's the next level?

John Dolmetsch, President - Business Information Group

We really need to be thinking about the future of our society and how this technology can change what we do and make our lives better rather than just providing access. The first step, of course, is that you have to provide the access. But unfortunately the access without the content is useless.

There are a lot of changes that cities need to plan adequately for. The wireless broadband access has certain functionalities that a traditional type of wired connection can never have if cities are prepared to take advantage of them. You can leverage things such as mobility and location based services to raise the quality and variety of services that you deliver. Vendors need to make sure that the cities understand those additional elements.

Judy Miller, President – Ninth Wave Media

Government needs to take an active role as a facilitator to bring this technology to the communities. Just how that role evolves will vary in every city. But if governments plan to stay focused on the goals of addressing the digital divide and improving neighborhoods, then whatever path they take to get there is worthwhile. Each city will have its own opponents and challenges, but they need to work through those to the best of their abilities.

I believe the next level of applications will result in a much higher degree of interactivity, particularly in the area of online learning. I see this being very beneficial for adults, helping them to complete their GED, or developing skills that enable them to get a job or advance in their current one. For students who are having trouble in school, this becomes another resource for them. But probably the initial focus will be on adults since the schools already have some good programs.

I also see an increase in the level of involvement of this technology in emergency preparedness. In New Orleans, the city's Web site was out of date at the time of the hurricane. Very few city Web sites are maintained well. A citywide WiFi network and community portals can be a great service during emergencies. Prior to and following Katrina, the portals could have remained up through co-located servers in other states so families and friends could try to get information into affected communities. As text messaging and other wireless services came back online, residents could have gotten word out to others.

As the technology becomes more sophisticated, businesses will be listed on community portals and this will help spur their economic development. Eventually it will be cool to have business get online to offer products and have purchases coming in from around the country, or even the world. I can see the day when the primary way that cash-constrained entrepreneurs will get started is to tap into mechanisms on portals to sell their products or services. They're not going to be opening a traditional storefront initially.

How is the technology scene evolving?

Cole Reinwand, VP Municipal Product Strategy and Marketing - Earthlink

It certainly seems to be the hot topic of the day, doesn't it? It's fashionable for cities right now to be considering municipal WiFi and justifiably so. The cable and DSL companies have not fulfilled their obligations in terms of providing broadband to certain constituents and doing so at a reasonable price. This isn't necessarily their fault, it's the economics at work here. But they also take advantage of their market position. When new technology like wireless comes along that offers superior economics and superior performance to enable affordable mobile broadband solutions, cities and private sector companies are going to start thinking about these things.

People recognize that we're on the cusp of a technology revolution. The same way that the telephone industry had a significant shift over the last 15 years to mobile voice, I think we're about the see the same thing happen with mobile data. This is going to enable all kinds of new usage paradigms and applications that have never been seen before. In two years, will municipalities still be leading the effort? I don't think so. I believe that the private sector will be driving the initiative. Municipalities will become the target of these companies. It's already happening. Earthlink is actively enticing municipalities to consider these kinds of networks, to understand the benefits.

Brad Tabaac, Owner – Friendly Pharmacy

The challenge with WiFi is the infrastructure behind it and making sure you don't create a wireless tower of Babble. I remember when PCs first came around I sold systems to drug stores that were all standalone machines. There was the rumor of one day having PCs talking to PCs, which wasn't possible then with the different PC operating systems. But that's where the technology is now, machines are talking to machines. It's solving the puzzle of moving data between different systems. WiFi is just a way of getting there.

If all the cities build WiFi systems, then the vendors will build the necessary applications so that all of the systems and devices connected to the network can talk to each other and use these applications. The wireless infrastructure will, in essence, drive the appropriate solutions. It's the chicken and the egg situation. Now I have some chickens, let's go get some eggs.

Jeff Arnold, Deputy Director of Legislative Affairs - National Association of Counties

As more and more devices become available and VoIP becomes more reliable, people will start to demand municipal broadband. These will be the drivers. The real question is will the spectrum be able to handle it. We forget about that sometime. At some point, the spectrum can only handle so much. We may find that we're going to bump up against these limitations because of so many devices.

Broadband over powerline [BBOP] is a way to get around the spectrum issue. It's a technology that Manassas, VA is using in a pilot which apparently is successful. This is a possible alternative for rural America as well. It's a little expensive in that you need a device at every transformer to be able to step down the signal to protect it from interference. Also, at the home you need a physical device to bring the connection to your computer. Unlike WiFi where you have mobility and a very small cost at the home, broadband over powerlines is targeted to home users. For people who need mobility, this may not work. They may use this for their computers in the home, but have regular wireless for traveling. In a rural community, most of the computers there are likely to be desktops, so in this situation broadband over powerlines makes sense.

What you're seeing is that we still need to have the demand side before this technology becomes affordable. There has to be some service coming in that can't come in any other way, but it has to be a service that people in the real world feel they can't live without. With that sort of demand developed, it makes this version of broadband worthwhile for the price point. If that price point significantly undercuts wireline broadband, then definitely people will flock to it.

Index